QC
522
.F37
2002

Chicago P...

BI

D0368624

icit

CHICAGO PUBLIC LIBRARY
BUSINESS / SCIENCE / TECHNOLOGY
400 S. ...
60605

DISCARD

# An Entertainment for Angels
## Electricity in the Enlightenment

Patricia Fara

**Revolutions in Science**
Series editor: Jon Turney

 **Columbia University Press**  New York

Columbia University Press

*Publishers Since 1893*

New York    Chichester, West Sussex

Copyright © 2002 Patricia Fara

First published by Icon Books Ltd., Duxford

All rights reserved

Library of Congress Cataloging-in-Publication Data

Fara, Patricia.

An entertainment for angels : electricity in the Enlightenment / Patricia Fara.

p. cm. — (Revolutions in science)

Originally published: Duxford, Cambridge : Icon Books, 2002.

Includes bibliographical references.

ISBN 0–231–13148–8 (cloth : alk. paper)

1. Electricity.  I. Title.  II. Series.

QC522.F37 2003

306.4'5—dc21    2003051565

Columbia University Press books are printed on

permanent and durable acid-free paper.

Printed in the United States of America

c 10 9 8 7 6 5 4 3 2 1

R0401798008

# Contents

CHICAGO PUBLIC LIBRARY
BUSINESS / SCIENCE / TECHNOLOGY
400 S. STATE ST.      60605

# Illustrations

# Dedication

For Helen and Katherine

# Acknowledgements

I should like to thank Simon Flynn and Jon Turney for their very helpful editorial suggestions. I am also extremely grateful to Jim Secord and Simon Schaffer, who have made it possible for me to write this book.

# Picture Acknowledgements

The illustrations on pages 14, 31, 45, 48, 53, 59, 80, 89, 126, 142, 149 and 157 are reproduced by permission of the Syndics of Cambridge University Library.

# Introduction

*What gave my Book the more sudden and general Celebrity, was the Success of one of its propos'd Experiments ... for drawing Lightning from the Clouds. This engag'd the public Attention every where.*

Benjamin Franklin, *Autobiography*, 1788

The poet Percy Bysshe Shelley was fascinated by scientific experiments. One of his friends described how he would pull out an electrical machine from the clutter on his desk, and charge himself up 'so that the fierce, crackling sparks flew forth; and ... his long, wild locks bristled and stood on end'. His eyes gleaming with enthusiasm, Shelley prophesied that science would find a way to keep poor people warm during the coldest winters. 'What a mighty instrument would electricity be', enthused Shelley, as he envisaged a future when 'by means of electrical kites we may draw down the lightning from heaven! ... an immense volume of electricity, the whole ammunition of a mighty thunderstorm; and this being directed to some point would there produce the most stupendous results.'[1]

Drawing 'Lightning from the Clouds': this visionary dream had been converted into reality by an American printer, Benjamin Franklin. Like Shelley, Franklin valued electricity for its potential benefits as well as its exciting effects. He was one of the eighteenth century's leading diplomats, internationally acclaimed for his enlightened attitude and democratic ideals. Yet he was also a prominent electrical researcher, famous for his numerous inventions and groundbreaking theories that had helped to establish the new science of electricity. Although he did not embark on his electrical career until he was forty, Franklin's scientific eminence paralleled his political reputation.

Electricity was the greatest scientific invention of the Enlightenment. Shelley was showing off his skills in 1810, but only a hundred years earlier, electrical science had not existed. Virtually the only way to produce electricity was by rubbing amber or a compliant cat, and even experts knew little that had not been familiar to the Greeks. Yet by the middle of the eighteenth century, electrical experiments were being performed all over Europe with new, powerful instruments that could produce, store and discharge static electricity.

Far from being an arcane preoccupation reserved for privileged intellectuals, electricity rapidly became a topic of conversation throughout society. Superlatives abounded: electricity was 'replete with wonders', 'as surprising as a miracle!', 'favourite object and pursuit of the age'. It was, one admirer exclaimed, 'an Entertainment for Angels, rather than for Men'.[2] Lecturers held their audiences spellbound as feathers jumped through the air, water jets glowed, and glasses of spirits were set aflame by the touch of a sword.

Many wealthy families bought their own apparatus, and aristocratic women produced miniature lightning flashes from their fingers and their whalebone petticoats, or titillated their admirers with a sensational – if rather painful – electric kiss. At the Hanoverian court, electric entertainment replaced dancing; in Edinburgh, myrtle trees were found to blossom earlier after a dose of electricity; while in Paris, electrified animals were found to lose weight (this was attributed to increased perspiration).

One early enthusiast was John Wesley, the Methodist preacher, who had learned about electricity from reading a book by Franklin. As Wesley travelled around England with his own electrical machine, he marvelled: 'What an amazing scene is here opened, for after-ages to improve upon.'[3] Journals like the *Gentleman's Magazine* echoed this enthusiasm, printing utopian suggestions of future applications for electricity. This new wonder power could, adherents claimed, be used for preserving hops and gunpowder, hatching eggs, or disinfecting bed linen; by impregnating food electrically, digestion might be improved – along with gout, the other major health preoccupation of valetudinarian gentlemen during the eighteenth century.

In his Oxford study, the electrified Shelley balanced on an insulated stool with his hair flying up, preaching about the most recent equipment to be invented – galvanic batteries that could produce electric current. Franklin and his contemporaries had only experimented with static electricity, but as the century drew to a close, Italian researchers discovered how to produce a flowing current. Shelley was enthralled. What further secrets of nature might be unlocked, he wondered, as science marched ever onwards? What new ways might be found of improving human welfare? Even in his wildest dreams, Shelley could not have known that he was indeed standing at the beginning of a new era, when electricity would come to run the world. It is hard for us to imagine life without electric power: Franklin and his Enlightenment colleagues made it possible.

# Part I
# Illuminations:
# The Light of Reason

*Why are the nations of the world so patient under despotism? Is it not because they are kept in darkness and want knowledge? Enlighten them and you will elevate them.*

Reverend Richard Price, *A Discourse on the Love of our Country*, 1789

For many years, Franklin's favourite portrait of himself was the one shown in Illustration 1, which had been painted in London in 1762 and engraved the following year. Following English conventions for portraying Enlightenment men of letters, Franklin is shown working in his study, his quill and paper prominently displayed to advertise his intellectual solidity. Clearly absorbed, he is listening to the electric bells behind his right shoulder, from which hang two electrified, mutually repellent cork balls. To his left, the traditional draped curtain is drawn

*Illustration 1*: 'Benjamin Franklin', mezzotint by Edward Fisher after Mason Chamberlin, 1763. (Wellcome Institute)

back to reveal an imaginary composite scene, in which the devastating effects of lightning are contrasted with the protective security afforded by Franklin's most famous invention, the lightning rod.

It was this picture that Franklin chose as a gift for his close friends and colleagues, and to impress remote government contacts in America. With the help of his son William, he sent out over one hundred copies, carefully rolling each one up in a protective tin case. Distributing pictures in this way enabled Franklin not only to boast about his electrical prestige, but also to solicit political support and consolidate personal friendships.

Beneath the engraving, the inscription 'B. Franklin of Philadelphia LLD FRS' makes it clear that Franklin was a distinguished scholar; moreover, it advertises that he was an American, the first President of the American Philosophical Society. As the colonies struggled for independence, nationalists enthusiastically vaunted specifically American achievements, hymning Franklin as the new nation's challenge to Isaac Newton. Poets gushed about this electrical expert whose political initiatives had guaranteed the political freedom necessary for scientific rationality to flourish:

*. . . we boast*
*A Franklin, prince of all philosophy,*
*A genius piercing as the electric fire,*
*Bright as the lightning's flash, explain'd so well*
*By him, the rival of Britannia's sage.–*

*This is the land of every joyous sound,*
*Of liberty and life, sweet liberty!*[4]

# 1 Interpretations

Franklin has become a hero not only for patriotic Americans, but also for scientists who want to tell dramatic stories about the discovery of electricity. Modern society depends on scientific and technologic achievements, but there are several different ways of describing how science has become so important.

Many historians envisage science as a progressive success story, a continuous march towards learning the truth about the natural world. They construct heroic models of the past in which exceptionally gifted men (and the occasional woman) build on the insights of their predecessors to make great advances. Achievements that may have taken years of effort are converted into picturesque adventures – Archimedes shouting 'Eureka!' from his bath, Isaac Newton sitting under an apple tree, James Watt watching a kettle boil. In these versions of science's history, Franklin appears as the great electrical champion who bravely conducted electricity down to earth by flying a kite.

Such memorable tales are appealing because they glamorise scientific research, but they present a misleading and over-simplified picture. Writers often smooth away conflicts, confusion and ambition, portraying scientists as disinterested searchers after truth, dedicated investigators who are unaffected by the normal demands of human life. Many scientists are, of course, genuinely concerned not only to discover more about the world, but also to bring about useful improvements. But they can also be competitive people who want to earn money, become famous and defeat their rivals. Despite the enthusiastic claims of Enlightenment experimenters, there was no easy upward path of progress, and research was fuelled by hostility as well as by curiosity.

Scientific experiments aren't always successful: people make mistakes, ignore results that later seem significant, or persuade themselves – and others – to adopt theories that turn out to be false. Just as importantly, science's accomplishments are not due to the single-handed efforts of a few geniuses. With the benefit of hindsight, it is relatively easy to pick out key individuals and events that seem to have directed the course of history. But singling them out means forgetting about the countless other projects that were being undertaken at the same time.

Franklin is central to any story of Enlightenment

electricity, but he was just one player in a complex tale of change that had many other participants – Joseph Priestley in England, Jean Nollet in France and Luigi Galvani in Italy, to name just a few. Franklin became interested in electricity only after he had read about other people's inventions, and his own innovations were adapted and improved by his successors. Even the story of his kite has been embroidered into a myth; in reality, some French experimenters were the first to show that lightning is electrical by drawing it down from the sky.

Science, politics and society were inextricably linked. When Franklin was eighty-one years old and helping to draft the American Constitution, an eminent English doctor sent him a flattering letter in which he wrote: 'Whilst I am writing to a Philosopher and a Friend, I can scarcely forget that I am also writing to the greatest Statesman of the present, or perhaps of any century, who spread the happy contagion of Liberty amongst his countrymen; and . . . deliver'd them from the house of bondage, and the scourge of oppression.'[5] Franklin's interest in electricity was rooted in this Enlightenment political ideology, and that is where this version of electricity's history will begin.

# 2    Electricity and Enlightenment

In the second half of the eighteenth century, there were only two English universities – Oxford and Cambridge. Both were expensive, closed to women, and better at training young men to become clergymen than scientific practitioners. Like Franklin himself, many people learnt about electricity by reading journal articles and attending lectures. One of the more successful popular scientific books was called *The Young Gentlemen's and Ladies Philosophy*, indicating that – unlike its heavier rivals – it was designed for women as well as for men.

*The Young Gentlemen's and Ladies Philosophy* uses dialogue, a traditional teaching method dating back to the Greeks. An elegantly dressed student, burdened with the appropriately classical-sounding name of Cleonicus, returns from university to play the superior role. He engages in long conversations with his envious sister Euphrosyne, who has been confined at home to study suitably feminine subjects such as drawing and dancing. In each chapter, the knowledgeable Cleonicus patronisingly explains the rudiments of a different branch of natural philosophy, the study of nature. By using terms so simple that even she can be expected to understand them, this artificial dialogue implicitly advertises that natural philosophy can be understood by everyone.

One day, after a lengthy session on the weather, Cleonicus agrees to explain how lightning and thunder are related to the exciting new science of electricity. Leading his sister into a shuttered room, he prepares to demonstrate a recent invention conveniently placed in their well-appointed home – an electrical machine (Illustration 2). Cleonicus is confident that with the help of this machine, he can dazzle his sister with dramatic effects of light and sound totally different from anything she has ever encountered before.

Euphrosyne: *But, pray, Cleonicus, can you, by any Experiments, shew me this Matter of Electricity; for otherwise it is talking to me in the Dark?*

Cleonicus: *Yes; come with me into this dark Room, and then you will view it in its proper Light.*

Euphrosyne: *Well! it is dark enough sure. What am I to see here?*

Cleonicus: *You will now see in the Dark what*

*Illustration 2*: 'A New Electrical Machine for the Table'. Benjamin Martin, *The Young Gentlemen's and Ladies Philosophy*, 2 vols, London, 1759–63, vol. 1, facing p. 301. (Cambridge University Library)

*you could not before perceive in the Light. I rub*
*the Tube with a Piece of Silk, and you see the*
*Sparks of Fire ... Flashes many Inches in*
*Length, and very much resembling the forked*
*Lightning of the Skies.*[6]

Writers of this period delighted in using elaborate puns, and these references to light and dark allude not only to the physical surroundings of this young couple, but also to their intellectual and spiritual illumination. French philosophers declared that they were living in the '*Siècle des Lumières*', the 'Century of Lights', and this was indeed the time when Europe's cities started to be brightly lit. London's citizens were among the first to install hanging glass lanterns that shone all night – one German prince even thought that the splendid display had been especially prepared to honour his visit. Cleonicus' electrical sparks and flashes were a smaller, home-spun version of the dramatic firework shows so beloved by wealthy party-givers.

Seeing was closely allied with knowing. Progressive thinkers often claimed that they were living in an enlightened age, when the bright flame of reason would dispel the dark clouds of ignorance and superstition. Yet this was also a profoundly Christian society, and religious imagery of divine light still pervaded literature and art. Through learning about electrical light, Cleonicus is arguing, Euphrosyne will gain an enlightened mind, the rational, secular equivalent of a soul inspired by the Light of God.

The period from roughly 1730 to 1790 is often loosely referred to as 'the Enlightenment'. During this period, many writers declared that rational thought would sweep away old errors and superstitions; reason would guarantee material and intellectual progress, as well as political liberty and freedom of thought. Faith in the Bible as the unique source of truth should, they insisted, be replaced by confidence in knowledge about the physical world gained by natural philosophers. Rather than relying on Aristotelian logic (which was often referred to as 'the teaching of the schools'), people should learn through experiment and reason. The Scottish philosopher David Hume recommended ritual book-burning:

> *If we take in our hand any volume of divinity or school metaphysics, for instance, let us ask,* Does it contain any abstract reasoning concerning quantity or number? *No.* Does it contain any experimental reasoning concerning matters of fact and existence? *No. Commit*

*it then to the flames; for it can contain nothing but sophistry and illusion.*[7]

One of the most famous Enlightenment philosophers was François-Marie Voltaire, who declared triumphantly that 'the spirit of the century . . . has destroyed all the prejudices with which society was afflicted: astrologers' predictions, false prodigies, false marvels, and superstitious customs'.[8] In accordance with Voltaire's boast, the Enlightenment came to mean the time when scientific, quantitative methods were successfully introduced. Often dubbed 'the Age of Reason', the Enlightenment is traditionally celebrated as the birth of modern civilisation, which took place primarily in France.

But it is time to rewrite the history of the Enlightenment. Rationality no longer seems such an unquestionable virtue, since spiritual and mythological approaches also have many insights to offer. There was no universal modernising spirit that swept across Europe, as though the electric light of reason had suddenly been switched on. Rather, the Enlightenment period was characterised by intellectual dissent, national differences and a persistent esteem for the past.

The roots of the Enlightenment lie further back than 1730, and England was also the home of enlightened values. As early as 1706, the third Earl of Shaftesbury proclaimed that 'there is a mighty Light which spreads its self over the world especially in those two free Nations of England and Holland; on whom the affairs of Europe now turn ... it is impossible but Letters and Knowledge must advance in greater Proportion than ever ... I wish the Establishment of an intire Philosophicall Liberty.'[9] When Voltaire was forced into exile in 1726, it was England that he chose as his place of refuge. Partly to criticise affairs in France, he lavishly praised English practices, politics and people. In particular, he singled out Newton, marvelling that such a dedicated scholar 'lived honoured by his compatriots and was buried like a king who had done well by his subjects'.[10]

Newton's work was vital for the claims made by enlightened philosophers that reason should reign supreme. Science is central to modern society, but the word 'scientist' was not even coined until 1833. Newton and his followers were known as 'natural philosophers' because they taught that knowledge could be gained by examining the natural world. They were trying to reconcile God's two books: the Bible and the Book of Nature. Deciphering God's blueprint for the world was, they argued, the surest route to learning more about God, as well as to improving human welfare.

Three centuries ago, natural philosophers did not enjoy the prestigious status of modern scientists. *Gulliver's Travels* is a particularly famous example of hostility towards scientific projects. In this complex satire, first published in 1726 (the year before Newton died), Jonathan Swift poked fun at inventors by portraying them foolishly trying to make sunbeams out of cucumbers, or build houses from the roof down. Faced with such attacks, natural philosophers struggled to justify themselves by demonstrating that their results were useful and reliable. Science had not been established as a professional career, and although many natural philosophers were independently wealthy, others needed to invent new ways of earning money from their activities. Innovative entrepreneurs gave lectures, wrote books, and devised instruments that would not only test their theories, but also attract paying spectators who, it was hoped, would admire their skills and recognise their control over the powers of nature.

Of all the branches of natural philosophy that would eventually become modern scientific disciplines, the most popular was the study of electricity. Many of the early investigations took place in England, but interest quickly spread abroad. Especially during the second half of the eighteenth century, electrical experiments dominated meetings at learned institutions such as the Royal Society in London and the Académie Royale des Sciences in Paris.

In *The Young Gentlemen's and Ladies Philosophy*, Cleonicus performed for his sister's benefit, but his readers provided a large virtual audience. He excelled in the theatrical techniques essential for Enlightenment lectures, which were entertaining performances as much as educational demonstrations. Delighting in arousing his sister's bewildered admiration, Cleonicus generated circles of fire, ignited gunpowder and illuminated fountains of water. Maliciously teasing her delicate sensibility, he wired up her dog and enrolled his servant to act as electrical executioner to a small bird.

But Cleonicus was also concerned with more serious matters, applying his skills to investigate both the nature of electricity and the uses to which it could be put. For instance, he replicated lightning inside their house, and explored the possible medical benefits of delivering shocks to different parts of the body. Although fictional, Cleonicus behaved like many experimental demonstrators who believed in teaching through entertainment. Enlightenment philosophers wanted not only to discover natural laws, but also to promote themselves by displaying their command of apparently inexplicable phenomena such as electricity.

One of England's most famous electrical experts was Joseph Priestley. He is now best known as the discoverer of oxygen, but this accolade would have infuriated him, since he supported a rival chemical system based on a weightless fluid called 'phlogiston'. Priestley would perhaps have preferred to be remembered as the founder of the fizzy drinks industry, launched when he sold the secret of his novel sparkling water to an enterprising colleague, Mr Schweppes.

Priestley gained such notoriety for championing radical political causes that his Birmingham house was destroyed by a patriotic mob in 1791. 'Electricity', he wrote, 'has one considerable advantage over most other branches of science, as it both furnishes matter of speculation for philosophers, and of entertainment for all persons promiscuously'.[11] Small fortunes could be earned, he continued, by those willing to study hard and lift themselves above the 'ignorant staring vulgar' who were so easily astounded by the wonders of electricity. Just as Cleonicus' electrical expertise elevated him above his sister, so too, enlightened philosophers regarded themselves as superior to the uninitiated.

Priestley was an extraordinarily versatile scholar, an expert on theology, ancient languages and education, as well as chemistry. In just over a year, he familiarised himself with modern electrical knowledge, and published a comprehensive book designed to enable ordinary people – including children – to learn about electricity through studying its history. This was no dry academic text, but a lively account that preached an Enlightenment belief: progress can be achieved through democratic education. Experimental discoveries, Priestley and his colleagues taught, offered the route for intelligence to rule the world. They hoped that scientific achievements would challenge existing systems of government that were based on birth rather than ability. Priestley provocatively warned that 'the English hierarchy (if there be anything unsound in its constitution) has equal reason to tremble even at an air pump, or an electrical machine'.[12]

Enlightenment philosophers hoped to gain authority over society by proving their dominion over nature. Electrical experiments provided dramatic evidence of their special powers. By the end of the eighteenth century, a turning-point often called 'the second scientific revolution', their efforts were bearing fruit. Modern scientific disciplines such as electricity, geology and magnetism were becoming consolidated. The Fellows of the Royal Society were successfully demonstrating the commercial and imperial value of their research, and so enjoyed a far higher status than in the time of *Gulliver's Travels*. Men of science were becoming respected members of society, and some of them were being paid for their work.

But such fundamental changes took place sporadically and unevenly: historical events simply do not neatly unroll in chronological order. There are, however, other ways of dividing up the past. Many modern historians prefer to think about themes rather than periods, using contrasting approaches to examine people and their activities from various angles. During the eighteenth century, when Franklin and his colleagues were dealing with static electricity, they had three basic goals: to devise new instruments; to find useful applications; and to formulate theoretical explanations for their discoveries. The next three chapters of this book explore these three aspects of Enlightenment electricity. The final chapter discusses the controversies surrounding the crowning climax of these Enlightenment investigations: the production of current electricity.

# Part II
# Shocking Inventions:
# Instruments

*A turkey is to be killed for . . . dinner by the* electrical shock, *and roasted by the* electrical jack, *before a fire kindled by the* electrified bottle: *when the healths of all the famous electricians in* England, Holland, France, *and* Germany *are to be drank in* electrified bumpers, *under the discharge of guns from the* electrical battery.

Benjamin Franklin, *Letter to Peter Collinson*, 1748

Electricity was the scientific vogue of the Enlightenment. Like other fads, interest erupted suddenly. During Newton's lifetime, electricity had attracted little interest, yet by the middle of the eighteenth century it was threatening to become an international obsession. Natural philosophers often boasted about modern achievements, and by 1767, only forty years after Newton's death, Priestley had already relegated Newton to the distant past. What, wondered Priestley, would Newton and the Greeks have made of 'the present race of electricians' who could perform so many entertaining experiments with their electrical machines?[13]

Priestley deliberately described himself as an electrician, a word first used in print by Franklin in 1751, in order to underline his modernness. As well as inventing new instruments, electrical experimenters were coining a specialised vocabulary, which can sometimes be confusing. 'Electrician' is just one of the novel words that have now passed into everyday usage, but have changed their meaning during the intervening centuries. Most importantly, when eighteenth-century writers referred to electricity, they meant static electricity: current electricity would not become useful until the nineteenth century. 'Battery' is another deceptive word: originally used for a group of guns, it gradually came to mean a collection of identical magnetic bars or electrical instruments linked together to accumulate their power. It was only in the early nineteenth century, after the discovery of current electricity, that it acquired its modern meaning.

The substances which displayed the greatest effects, and which interested electricians the most, were what we call 'insulators'. '*Electron*' is the Greek for amber, whose ability to attract small pieces of straw or feathers after it has been rubbed had been known for thousands of years. At the beginning of the seventeenth century, substances that can remain charged, such as glass, amber and many gemstones, were christened 'electrics'. In contrast, since metals and other conductors transmit any charge – or 'electrical virtue', as it was often known – they were labelled 'non-electrics'. Iron gun barrels, lead-lined flasks and gold leaves became essential components of electrical apparatus. Water was another non-electric that came to play a vital role in storing and transporting large electric charges. For many experimenters, the most interesting non-electrics were people and animals (either living or dead), followed closely by growing plants.

Unlike related topics such as magnetism and optics, very little research was carried out into electrical effects before the beginning of the eighteenth century. Earlier natural philosophers had performed a few exploratory experiments, but they had discovered little that was not familiar to the Greeks. Partly because they were more interested in magnetism, their examinations were almost entirely limited to the attractive and repulsive properties of electrics, and they knew nothing of the flashes of light and crackling sounds that so intrigued Enlightenment philosophers.

So what sparked off this sudden interest in electricity (this pun may seem painfully weak, but it is typical of those that abound in eighteenth-century texts)? Science is often said to advance when new theoretical ideas are put forward, but in this case it was definitely the instruments that came first. Indeed, one of the most important – the Leyden jar – should not even have worked at all according to the prevailing theories. Experimenters often had no viable explanations for the strange phenomena they were witnessing, but they developed many new pieces of equipment that displayed novel and dramatic effects. At the time, they had no idea that the entertaining performances they devised for their audiences would eventually result in the electro-magnetic networks that permeate modern society.

Enlightenment electricity stemmed partly from the casual interest of seventeenth-century natural philosophers in amber, sealing-wax and other electrics. But there was no clear path of development. An equally important origin for the Enlightenment obsession with electrical sparks and flashes was the awed wonder aroused by luminescent materials, glowing substances that seemed to belong to magic rather than science. Yet it took more than fifty years for experimenters to make the connection between these two fields.

It is only in retrospect that we can construct a coherent history of electricity that apparently leads to our present day, a distant future undreamt of even by the most optimistic of Enlightenment utopians. Three new instruments were particularly vital for static electricity during the eighteenth century. Air-pumps, invented to create a vacuum, led to the accidental discovery of electric glows. Electrical machines, based on a rotating glass globe, propagated the Enlightenment fervour for electricity. And Leyden jars, introduced at mid-century, enabled electric charge to be stored and transported from one place to another.

# 1 Robert Boyle and the Air-pump

In the 1660s, the Fellows of the newly-founded Royal Society regarded Robert Boyle's air-pump as their most splendid achievement (Illustration 3). As the diagram makes clear, the air-pump was a feat of technical engineering, designed to evacuate the large glass globe at the top. At its base, the globe is connected through a stop-cock to a brass cylinder, inside which a piston is moved up and down by turning the handle to remove the air. The stopper at the top of the globe enables objects to be placed inside, in order to see how they behave in this airless environment. For example, a ringing bell gradually becomes silent as the air is withdrawn, demonstrating that sound cannot travel through a vacuum.

Boyle (1627–91) was a wealthy Irish aristocrat with sufficient money and time to perform a huge variety of experiments. Most famous now for his law describing the behaviour of gases, he was admired by his contemporaries for his religious observance, his new ideas about chemistry, and his air-pump. He first started working on air-pumps in Oxford, where he

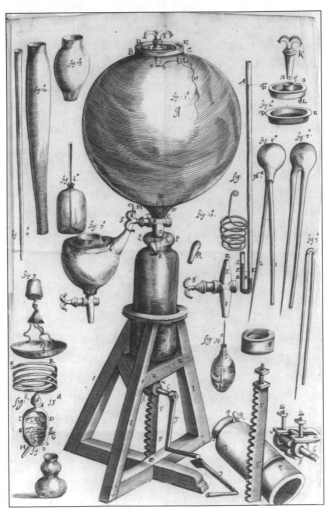

*Illustration 3*: Robert Boyle's first air-pump. Robert Boyle, *New Experiments Physico-Mechanicall Touching the Spring of the Air*, Oxford, 1660. (Cambridge University Library)

belonged to an active circle of experimenters which included the architect Christopher Wren. Plagued by ill-health and lovingly cared for by his sister, Boyle was a rich gentleman with his own private laboratory. Although he ensured that his name was attached to the pump, much of the hard labour (and probably also the creative work) was carried out not by Boyle, but by the brilliant yet impoverished inventor Robert Hooke. In addition, there must have been a team of backroom technicians – now often called 'invisible assistants' – about whom we know virtually nothing.

The air-pump represented a conceptual as well as a mechanical triumph. Following Newton's ideas, we have become used to visualising interplanetary space as literally that – empty space. But in the seventeenth century, many natural philosophers believed the universe to be completely filled by tiny, invisible particles, so that establishing a vacuum was, they thought, intrinsically impossible. And even when Boyle did manage to convince them that his glass globe was virtually empty, his critics protested that it did not make sense to study nature by creating such an artificial situation. By insisting that experiments with his air-pump provided valid evidence about the world, Boyle established a new route to obtaining knowledge.

All over Europe, natural philosophers devised countless experiments to be performed with air-pumps. For the first time, they could investigate the properties of air by examining what happened in its absence. Boyle himself carried out many experiments: he compared the results of burning different substances on a red hot plate inside the globe, showed that candle flames died away as the pump was operated, and used small birds and animals to demonstrate that air is needed to support life.

Although he did examine the behaviour of amber and other electrics, Boyle was far more interested in his barometer experiments, which he regarded as 'the principal fruit I promised myself from our engine [the air-pump]'.[14] To resolve arguments about what was in the space above the mercury, Boyle placed a barometer tube inside his glass globe and watched what happened. Just as he had predicted, the level of mercury descended as the air was removed, thus confirming experimentally that there was a vacuum inside the tube. Barometer tubes were subsequently to prove very significant in electricity's history.

The air-pump became a potent symbol of the new scientific methodology promoted by the Royal Society. A century later, the Derby artist Joseph Wright painted what is now his most famous picture, 'An Experiment on a Bird in the Air Pump'.

Reproduced on numerous modern dust-jackets, this painting still provides a powerful yet enigmatic comment on the rewards of scientific experimentation. The canvas is dominated by a red-robed and slightly dishevelled experimenter, whose raised eyebrows and beckoning hand invite the viewer to pass judgement. Should he let air into the globe, so that the white cockatoo trapped inside can live, or should he continue his investigations, sacrificing the bird to the cause of research? Wright also visually poses another question. Should we be enlightened by the natural light glimpsed through a small window in one corner, or allow ourselves to be overwhelmed by the centrally placed yet artificial light source that illuminates this scene, an eerily glowing goblet containing a skull?

This same contrast between natural and artificial light appears in another of Wright's paintings, 'The Alchymist, in Search of the Philosopher's Stone, Discovers Phosphorus, and prays for the successful Conclusion of his operation, as was the custom of the Ancient Chymical Astrologers', first exhibited in 1771 (Illustration 4). Kneeling before a stoppered flask that irradiates his dark and cluttered laboratory, the hopeful alchemist raises his eyes to heaven for guidance. Wright's composition deliberately recalls traditional religious imagery, since the

*Illustration 4*: Joseph Wright of Derby, 'The Alchymist, in Search of the Philosopher's Stone, Discovers Phosphorus, and prays for the successful Conclusion of his operation, as was the custom of the Ancient Chymical Astrologers', 1795. (Derby Museums and Art Gallery)

experiment's shining chemical light ambiguously hovers between secular and divine illumination. Concealed within his Platonic cave, the self-seeking philosopher ignores God's moon glimmering through the Gothic window, and instead relies on man-made enlightenment as he vainly strives towards his unrealisable goal.

Wright was well versed in modern chemical techniques, and in tune with the progressive aspirations of Priestley and his associates. He intended this picture not as a literal illustration of alchemical practices, but as an Enlightenment commentary on the validity of different approaches to knowledge. By focusing on phosphorus, Wright depicted research that had been taking place in England about a hundred years earlier, when the air-pump's success was being consolidated.

This had been an anxious time for natural philosophers, since they urgently needed to distance themselves from magicians, astrologers and other arcane practitioners. As Boyle and his contemporaries investigated combustion, they became fascinated by substances collectively called 'phosphoruses'. Mysteriously, these emitted light as though they were burning, yet remained cold. Along with rotting wood, putrefying fish, glow-worms and fireflies, this group of phosphoruses included a wonderful new mineral from Bologna, which glimmered in the dark after it had been exposed to light. One visitor to England scattered fragments on Boyle's carpet, where they twinkled like stars, and wrote '*DOMINI*' (Latin for 'of God') in gleaming letters on a piece of paper. German alchemists claimed that this 'Bolonian stone' attracted light because of its occult sympathetic bonding with the sun, but the Fellows of the Royal Society were determined to find a rational explanation. Uncovering the secrets of this substance – what we now call phosphorus – would, they believed, help to establish a firm distinction between experimental philosophers and magicians.

Another type of luminescence that puzzled investigators was the purple glow that sometimes appeared in the space above the mercury in a barometer's U-shaped tube. First discovered when a tube was accidentally shaken, it was this 'mercurial phosphorescence', or 'barometric glow', that brought together two separate fields of research – the air-pump and phosphorescence. Although no one realised it at the time, some theatrical demonstrations of this mysterious light, initially devised to entertain the leisured members of the Royal Society at the beginning of the eighteenth century, would eventually result in the electrical machine that Cleonicus used to impress his sister Euphrosyne.

# 2  Francis Hauksbee and the Electrical Machine

In stark contrast with Boyle, Francis Hauksbee (*c.* 1666–1713) was no landed gentleman, but a draper who was already in his late thirties when he made his first appearance at the Royal Society. His origins remain obscure, but he may have learnt about air-pumps by working as one of Boyle's behind-the-scenes assistants. Somehow he managed to obtain Newton's approval, and, despite his lack of formal education, was rapidly promoted from his post of experimental assistant to become a Fellow of the Royal Society.

Hauksbee was keen to improve the air-pump's design, but he also needed to justify his salary. He was employed to devise dramatic weekly demonstrations that would sustain the interest of Fellows who were not committed experimenters. By examining mercurial phosphorus, he could fulfil both his objectives. Working over several months, Hauksbee managed to combine showmanship with systematic research.

In one spectacular performance, he found that he

could obtain mercurial phosphorescence without the barometer. Hauksbee produced 'a Shower of Fire' by passing a stream of air through a pool of mercury inside an air-pump's globe. Repeating similar trials under different conditions enabled him simultaneously to satisfy the Fellows' appetite for wonder, and to establish that a vacuum was not essential for the effect. Next he discovered that not even the mercury was necessary, since he could obtain a glow by rubbing together two objects inside the air-pump. The machine that he devised to perform this tricky task was a masterpiece of mechanical ingenuity.

Hauksbee mounted an evacuated glass globe of about 20 cm diameter on to a spindle so that it could be rapidly rotated. By pressing the palm of his hand against the spinning glass, he produced a purple light brilliant enough for him to read by. Even watching a modern replication of this experiment is an exciting experience. As the wheel turns noisily in a darkened room, the globe is gradually filled with an eerie purple glow laced by brighter streaks. How much more stunning it must have been in the early eighteenth century, when such an artificial light had never before been seen.

It was only at this stage that Newton and Hauksbee linked these results with electricity – and even then, they were more interested in the optical effects of glass. Conjecturing that Hauksbee might have been forcing small particles of light out of the glass, they wondered hopefully whether this research might provide a bridge between Newton's two great fields of interest, gravity and optics. Hauksbee's first move was to equip himself with a new instrument large enough for his weekly performances – a hollow glass tube about 2.5 cm across and 75 cm long. By rubbing this electric wand, he could produce a glow and make small pieces of brass dance in the air as if by magic. Hauksbee also contrived ingenious arrangements of hanging threads and whirling globes to detect the movement of tiny, invisible particles called 'effluvia', which he thought might be responsible for the effects that he observed.

However, faced with apparently contradictory results, Hauksbee set these investigations aside, and turned to other ways of vindicating Newton's ideas through spectacular experiments. Since he died a few years later, he never realised that his tube and rotating globe would become standard items of equipment for Enlightenment electricians. Hauksbee had gradually developed specialised equipment in order to explore glass's strange optical effects. His one-off device was taken out of the restricted environment of the Royal Society, and converted into a powerful machine familiar to every fashionable consumer of electrical entertainment. The early stages in this transition were due to an enterprising and determined silk dyer from Canterbury called Stephen Gray.

# 3 Stephen Gray and the Charity Boy

Enlightenment educators preached that personal self-improvement would lead to economic and social benefits for the whole of society. Through publishing the *Philosophical Transactions*, the Royal Society hoped not only to solicit reports of exciting discoveries, but also to enlighten less privileged readers such as Gray, who first learnt about Hauksbee's inventions from a borrowed copy of the Society's journal. Gray (1666–1736) belonged to a well-established family of tradesmen, and like many other provincial artisans, believed that acquiring knowledge would go hand in hand with gaining wealth and social status.

Based in Canterbury, Gray found it hard to stimulate metropolitan interest in his electrical suggestions. Already self-taught in microscopy and astronomy, Gray decided to better himself by moving to London and learning more about electricity. Because he had suffered an injury in his dyeing work, he gained a position in the Charter-house, a charity school in which he could pursue his electrical research. Now centrally located in the capital, he eventually managed to overcome the disadvantages of his origins.

When trying to repeat Hauksbee's experiments with his glass tube, Gray had noticed that feathers were drawn to the cork at each end, rather than to the glass itself. Other experimenters were concentrating on attraction and repulsion, but Gray set out to explore how electric charge – often called 'Vertue' – could be transmitted from one place to another. Concluding that the tube's electrical virtue had been communicated to the cork, Gray tried to transmit it to other objects over greater distances. Starting with fishing-rods, he used longer and longer lengths of thread and wire to electrify coins, vegetables and even his kettle. However, as he got more ambitious, he outgrew his small London room, and migrated to the country estates of Fellows willing to accommodate his apparatus. Gray successfully set up 'lines of communication' that stretched over 250 metres, using them to affect a diverse assortment of objects, including soap bubbles, a sirloin of beef, a red-hot poker, a map, an umbrella – and even a small boy.

After more than twenty years of struggle, Gray finally gained recognition in 1730, when he was elected a Fellow of the Royal Society. For the next few years, learned audiences flocked to the

Charterhouse to witness the exciting experiments that he performed for their entertainment. Although these fashionable ladies and gentlemen ignored Gray's theoretical suggestions, they were particularly captivated by his *pièce de résistance*, the hanging boy (Illustration 5). Taking full advantage of available research materials, Gray suspended one of the charity school pupils by two clothes lines from the ceiling, and electrified him with a glass tube. To the audience's delight, the electrified boy attracted feathers, brass flakes and other light objects with his hands.

Illustration 5 is taken from a book by Jean Nollet, one of the earliest French natural philosophers to popularise Newton's ideas. Originally trained for the priesthood, Nollet – like Gray and Franklin – came from a humble background, and earned his living by giving experimental demonstrations in Paris. Nollet and other electrical researchers repeated Gray's performance, often exploiting its dramatic potentials. Here the electrician resembles a conjurer wielding his wand, demonstrating his apparently magical powers by lightly touching the boy's body. Beneath the boy's right hand, a textbook, the conventional source of knowledge, lies open yet discarded. Fascinated onlookers cluster round: they have to learn that if they draw too near, they will be

*Illustration 5*: The hanging boy. J.A. Nollet, *Essai sur l'éléctricité des corps*, Paris, 1746, frontispiece. (Cambridge University Library)

startled by painful crackling flashes, a particularly effective trick in the dark.

Gray and his colleagues electrified their glass rods by rubbing them with pads of leather or wool. In the 1740s, experimenters discovered that they could achieve more striking results by making new versions of Hauksbee's machine, such as the table-top version in Illustration 2. As the experimental assistant turns the handle, the horizontal metal rod, which is called a 'prime conductor', transmits the electricity generated by the rubbed globe. Electricians throughout Europe devised variations of the basic model. They used swords or gun barrels to act as prime conductors, and installed leather-covered rubbing cushions on springs to free the experimenter's hands. Other ingenious features included linked globes, protective flaps of silk impregnated with beeswax, and metal collecting combs.

Entrepreneurial experimenters concocted elaborate devices to attract spectators. In one performance, fashionable women sitting comfortably on a swinging seat grasped the metal conductor of a large machine, so that sparks would fly when they were touched. Candles were relit, alcohol was set aflame with electrical water, and unsuspecting dinner guests were shocked by electrified cutlery. Some suspended boys were charged up by having the soles of their feet pressed against the spinning glass globe, so that they themselves became the prime conductor, transmitting electricity to people standing on insulating stools made of pitch and resin.

There was no clear boundary between electrical performers who acted as magicians, and lecturers who delivered more serious information about the powers of nature. Even the demonstrations at the Royal Society were designed to amuse as well as instruct the Fellows: '[A]ll the variety of Tricks, I call them, rather than experiment[s] on Electricity', grumbled one elderly electrician.[15] As electricity became more fashionable, London instrument makers started to market a range of demonstration equipment, which they sold to private customers for parlour performances as well as to the growing band of travelling lecturers.

Electrical entertainment was starting to offer business opportunities. Benjamin Rackstrow, who specialised in making busts and garden ornaments for London's aspiring classes, decided to diversify into electrical lecturing. According to his advertising pamphlet, distinguished visitors included the Duke of Cumberland, foreign ambassadors and the president of the Royal Society. To attract further customers, Rackstrow installed his own version of the beatification (Illustration 6), an ordeal that had

1. *The Glass Crown.* 2. *Stop Cock.* 3333. *Points turnd downwards.* 4. *Brass Bottom.* 55. *Electric plate & Wire.*

*Illustration 6*: Benjamin Rackstrow's beatification. Frontispiece of: Benjamin Rackstrow, *Miscellaneous Observations, Together with a Collection of Experiments on Electricity*, London, 1748. (Cambridge University Library)

originally been inflicted on young boys by a German experimenter. Standing on resin blocks, each boy had been electrified until his entire body was bathed in a glowing light resembling a saint's halo.

For his shop window display, Rackstrow made a glass crown, with a brass plate at the bottom, and a tin lid carrying a stop-cock so that the space inside the crown could be evacuated with an air-pump. Seating his victim in a chair with an adjustable back, Rackstrow electrified a suspended plate so that 'a continued stream of Fire would appear between the Plate and the Crown, and the Crown look luminous, as if almost fill'd with fire, by numberless rays of light darting in different forms from top to bottom in a glorious manner'.[16]

Pious natural philosophers sneered at such Enlightenment entrepreneurship, protesting that it was immoral to gain financially from a natural wonder created by God. But lecturers were competing with conjurers, singers and other entertainers. They devised compelling electrical performances that blended dramatic display with instructive education, and tried to supplement their profits by publishing chatty books, such as the one featuring Cleonicus and Euphrosyne. As a young man, John Smeaton (who many years later became famous for engineering the Eddystone lighthouse) built his own electrical machine and defended his attempts to recoup this expenditure through showmanship.

'I don't take it yt shewing ye wonders of Electricity for Money is much more commendable than ye shewing any other strange sight or Curiosity for ye same end', he acknowledged. On the other hand, he continued, 'if £200 could be got by a worthy employment in yt way I don't see wheres ye harm as there is no fraud nor Dishonesty [in] it'.[17]

As Smeaton implied, experimental philosophers were as intrigued as everyone else by electricity's dazzling effects, and making money from the wonders of electricity did not necessarily preclude more serious investigations. In the absence of any professional career structure for men of science, these lecturers, instrument makers and writers needed money to finance their research. Although they each had their own interests at heart, men like Hauksbee, Gray and Priestley collectively helped to establish the national importance of experiments being carried out in élite institutions like the Royal Society. They transformed natural philosophy from an arcane interest into a public science that would benefit the entire nation. These entrepreneurial philosophers did not only invent new instruments – they also unwittingly participated in the larger, cumulative project of inventing modern science.

# 4 Pieter van Musschenbroek and the Leyden Jar

The Leyden jar was the most influential electrical invention of the eighteenth century. As well as providing larger charges than had previously been possible, for the first time it enabled electric charge to be stored and conveniently moved from one place to another. However, like so many scientific discoveries, the Leyden jar arose not from any great theoretical insights, but as the unexpected consequence of trial and error operations.

Even the instrument's name seems accidental, since it was first created not in Leyden, but in Kammin, a remote Pomeranian town now in Poland. In 1745, one of the cathedral deans, Ewald von Kleist, was experimenting with his electrical machine. Using convenient tools such as an old medicine bottle and a nail, he claimed that he could produce shocks strong enough to knock young children off their feet. Unfortunately for his posthumous reputation, von Kleist kept his methods so secret that no one could replicate his results. Instead, the accolade of inventor went to Pieter van Musschenbroek (1692–1761), a professor at the University of Leyden, who the following year independently stumbled upon the instrument that would revolutionise electrical research. For the first time, electric charge could be stored and transported in what we would call a 'condenser' or a 'capacitor'.

According to Voltaire, Musschenbroek was 'naivety itself and loved Truth with a child's openness'. After unintentionally stunning himself with a massive electric charge, he wrote a long letter to a close colleague in Paris about his 'new but terrible experiment, which I advise you never to try yourself, nor would I, who have experienced it, and survived by the grace of God, do it again for all the kingdom of France'. This disingenuously phrased advice was, of course, a direct invitation to admire Musschenbroek's ingenuity, and he helpfully supplied detailed instructions on how to build his new device, even specifying that German rather than Dutch or English glass should be used.[18] While Musschenbroek continued to refine his literally shocking experiments at Leyden, then one of the world's leading universities, Parisian researchers immediately started to make their own Leyden jars.

The lower part of Illustration 7 shows the first published drawing of Musschenbroek's experiment. It appeared in a book produced later the same year

*Illustration 7*: The Leyden experiment, and (above) the flow of electrical matter. J.A. Nollet, *Essai sur l'éléctricité des corps*, Paris, 1746, facing p. 216. (Cambridge University Library)

by Nollet, who named the Leyden jar in direct tribute to Musschenbroek. At the right, an electrical machine is charging up a horizontal metal rod or gun barrel, which, suspended by silken threads, acts as the prime conductor. At its other end, a wire dips down into the Leyden jar, a glass flask containing water, which stores the charge being transmitted to it. Holding the jar in one hand, the experimenter feels nothing. But when he touches the wire or the prime conductor with his other hand, and the accumulated charge surges through his body, he immediately feels a strong pain in his arms and chest, and runs the risk of nosebleeds, temporary paralysis, convulsions and prolonged dizziness.

Unlike many descriptions of electrical instruments, Nollet's illustration makes it absolutely clear that at least two experimenters are required: a trained helper operates the electrical machine, while the chief investigator controls the Leyden jar. Vital assistants were normally invisible, but one also appears in Illustration 2 (on page 14), which shows Cleonicus' servant John turning the handle of the electrical machine. John, only briefly referred to in the text and apparently ignored by his employers, is not only cut off by the edge of the picture, he is also drawn disproportionately small. In contrast, Nollet and his assistant seem to be working as a team. Still more surprisingly, she is a woman! Women in France were far more able to participate in intellectual activities than in England, where men clubbed together exclusively in their learned societies, coffee houses and masonic lodges. But in Paris, influential women headed *salons* attended by men as well as women, and although far more cloistered than now, were relatively free to engage in academic debates. Paradoxically, they were to lose this enlightened liberty after the Revolution, when it became fashionable for women to adopt a strong maternal role.

Replicating Musschenbroek's experiment was a hazardous yet compulsive exercise, and attracted people from many walks of life. One visiting English clergyman paid to be shocked in Paris, noting smugly that the experienced operator flinched from demonstrating the painful effects. Writing from Leipzig, a classics professor reported his personal experiences to the Royal Society in London, knowing that publication in the *Philosophical Transactions* would ensure international coverage. 'I found great Convulsions by it in my Body. It put my Blood into great Agitation; so that I was afraid of an ardent Fever; and was obliged to use refrigerating Medicines. I felt a Heaviness in my Head, as if I had a Stone lying upon it. It gave me twice a Bleeding at my Nose, to which I am not inclined '[19] Despite these alarming symptoms, he subjected his wife to similar treatment, weakening her so that she could hardly walk.

As with the electrical machine, entertainment and

investigative research were closely intertwined. Careful experiments soon established that instead of using water, the glass could be lined inside and out with lead foil. Provided that the wire at the top was not touched, a Leyden jar would hold its charge for hours or even days, and by connecting ten, twenty, or even a hundred jars together in batteries, the shock could be increased still further. One of these batteries, now on display at the Boerhaave museum in Leyden, carries an adjustable dial with four settings: detonating cannon, altering a compass needle, killing small animals, and melting wire.

Another approach was to explore how effectively the shock from a Leyden jar could be transferred from one place to another. Nollet was the acknowledged expert at arranging spectacular demonstrations involving long chains of people holding hands. The first person in the line held a Leyden jar, and when the last one touched its prime conductor, they all jumped up into the air one after the other as the charge passed along them. At Versailles, Nollet entertained the king with 180 leaping soldiers; he achieved a similar feat with 200 Carthusian monks at their monastery, and later reached a record of over 600 people in his living chain. Other experimenters extended these trials by connecting people together with long wires, and electric discharges were soon being transmitted around the Tuileries gardens, across Westminster Bridge, and through the River Skuylkil in North America, where Franklin's celebratory plans included roasting an electrocuted turkey on an electrically rotating spit.

Events did not always proceed smoothly. On one occasion, a routine demonstration with sixty students was unexpectedly interrupted when the shock failed to pass beyond the sixth person in the chain. Since it was well known that women were more susceptible to electrification than men, rumours began to circulate that the young man forming the obstacle was – as the experimenter delicately expressed it – 'not endowed with everything that constitutes the distinctive character of a man'. To test this hypothesis, the experiment was repeated with three famous counter tenors from the royal choir who had been surgically operated upon, but they successfully transmitted the shock to their neighbours. Could there be a difference in electrical susceptibility between 'men who have been mutilated by Art and those towards whom Nature has shown herself to be a hard-hearted mother'? The debates dragged futilely on, until it was eventually realised that the unfortunate culprits who impeded the charge's progress were always standing on a patch of wet ground, which was conducting the electricity down and away from the more resistant human chain.[20]

Paris was the focus of activity for the first trials of the new Leyden jar, but within a few months its possibilities were being enthusiastically explored in England, Italy, Germany and the British colonies of America. The journals of learned societies were dominated by reports of experiments and medical investigations, and by the end of the eighteenth century, electricity formed a major component of every course on natural philosophy, as well as providing a good source of home entertainment for middle-class families. As just one indication of electricity's widespread appeal, the article in the 1797 *Encyclopædia Britannica* ran to over 120 pages, and included 86 diagrams.

Illustration 8 shows just one of several plates in a book by Adam Walker, a well-known lecturer in Britain at the end of the eighteenth century. His most famous pupil was Shelley, who as a schoolboy had been inspired by Walker's scientific lessons. Walker's *System of Familiar Philosophy* exemplifies the Enlightenment concept of rational entertainment for home readers. Serious discussions about the nature of electric charge are mixed with amusing demonstrations, while the explanations are verbal rather than mathematical, couched in relatively simple

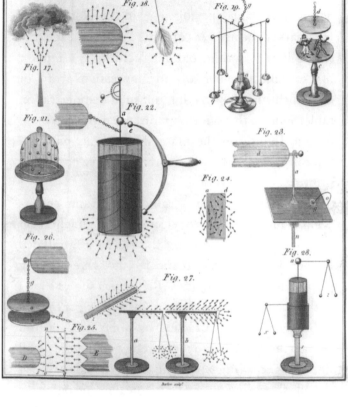

*Illustration* 8: Adam Walker's electrical experiments.
Adam Walker, *A System of Familiar Philosophy: in Twelve Lectures; Being the Course Usually Read by Mr A. Walker*, London, 1799, Electricity Plate II. (Cambridge University Library)

terms to make them accessible to those without the benefits of a good education.

Walker originally came from Manchester, but his lecture tours were so successful that he achieved national acclaim, performing in London at the Haymarket theatre. A caricature by James Gillray shows him as a thin, balding lecturer in a modest black jacket, owner of an impressive array of equipment that he is laying out on a demonstration table. Some of this same apparatus features in a more technical guise on the plate shown here, which is typically crammed with diagrams to save paper and printing costs.

The small arrows in Walker's diagrams correspond to his discussions of electric charge. Light relief is deliberately provided by the electrically inflated feathers in his Figures 16 and 18 (the corresponding picture in the *Encyclopædia Britannica* shows a man with his hair standing out at right angles); at the top right, a brass plate attached by a metal chain to a prime conductor can be lowered to make the small figures dance. Other popular tricks included the set of bells shown in Figure 19 and the dancing pith balls of Figure 21, which alternately sprang apart and came together as a charged glass tube was brought near and withdrawn. Walker and other lecturers used these devices to capture attention, and also to illustrate theoretical points. The largest central diagram (Figure 22) illustrates an experiment with a Leyden jar.

As a provincial self-made man, Walker was regarded with suspicion by snooty Londoners: the novelist Fanny Burney was perturbed to find herself sharing a dinner table with such a 'vulgar' conversationalist. Coming from outside the established institutional framework, men like Walker challenged the structure of English society, and contributed to the establishment of electricity as a new scientific discipline in the early nineteenth century. Their achievements depended not on new scientific theories, but on instruments that could produce dramatic effects – especially the electrical machine and the Leyden jar.

# Part III
# Lightning Cures:
# Applications

> ... *there are two great advantages derived from Electricity; the one is a defence against the direful effects of lightning, and the other a remedy for many disorders incident to the human body.*
>
> Tiberius Cavallo, *A Complete Treatise on Electricity, in Theory and Practice*, 1795

Modern society is so dependent on electricity that it is hard to imagine the enormous excitement aroused by the first electrical gadgets. Artificial spiders, dancing balls, flaming books, tinkling bells and scintillating pledges of *amour* provided intense fascination for adults as well as children. Wealthy shoppers spent large sums of money on experimental apparatus that provided entertainment as well as instruction. London's instrument makers were widely acknowledged to be the best in the world, and they competed intensively against each other for this new middle-class income.

To satisfy – and create – this demand for rational amusement, electrical inventors devised complicated devices. One favourite parlour trick was the king's portrait, whose wired-up frame delivered a sharp shock to any anti-royalist dinner-party victim who yielded to the temptation of dislodging the royal crown. For a highly charged picture, surmised Franklin, 'the consequence might perhaps be as fatal as that of high treason'. However, he dispassionately remarked, he had not yet himself managed to use electricity for killing anything larger than a hen.[21]

Franklin's comment indicates how greatly eighteenth-century attitudes towards pain, death and humour differed from our own. In 1789, two caricatures were published showing a pseudo-nymous 'Count Fig' sprawling on the pavement, stunned after pressing an electrified door bell. This incident presumably aroused great mirth:

*The Count attacked the Inchanted Wire*
*Unconscious of the latant Fire.*
*Which hurled him prostrate on the Stones*
*Screaming aloud my bones – my Bones.*[22]

While this may now seem a practical joke in

singularly bad taste, a similarly sadistic approach was also adopted by more serious investigators. William Watson, London's leading electrician in the middle of the century, gave elaborate instructions for surprising an unsuspecting experimental subject with his 'electrical Mine', a spot on the carpet concealing wires to some strong Leyden jars. Watson coolly recorded the varying degrees of severe shock his visitors experienced under different conditions.[23]

Critics questioned the value of these experiments, voicing their doubts about the ambitious promises that Watson and other electricians were making of electricity's potential benefits. Watson became the target of one of the erotic lampoons that flourished in the eighteenth century. Supposedly written by Paddy Strong-Cock, it took the form of a book-length letter called *Teague-root Display'd: Being some Useful and Important Discoveries Tending to Illustrate the Doctrine of Electricity*. Using savage sexual satire, the anonymous author mocked the aims and achievements of W__M W__N and his fellow 'learned Luminaries'. Samuel Johnson expressed his own reservations about the Royal Society's activities with more discreet sarcasm: 'I have fallen eleven Times speechless under the Shock of Electricity; I have twice dislocated my Limbs, and once fractured my skull in essaying to fly; and four

Times endangered my Life by submitting to the Transfusion of Blood.'[24]

Despite – or perhaps provoked by – these cynical commentators, many electrical innovators concentrated on finding beneficial applications for their powerful Leyden jars. Two major uses were developed for electricity during the eighteenth century. The most famous is the lightning rod, in the development of which Franklin played a central role. But Franklin was also interested in Enlightenment electricity's other great application: medicine. Electrical therapy may now seem a dubious practice, but it was then taken very seriously.

Depending on their personalities and financial circumstances, researchers were motivated by fame, money and curiosity, as well as by a genuine desire to help other people and to make the world a better place to live. Collectively, their efforts validated electrical research by establishing its usefulness, thus promising to fulfil one of natural philosophy's Enlightenment ideals. Franklin smugly preached that even if 'there is no other Use discover'd of Electricity, this, however, is something considerable, that it may *help to make a vain Man humble*'.[25]

# 1    Benjamin Franklin

In Franklin's summary of his *Autobiography*, electricity is breathlessly squashed in between other pressing activities: '. . . Propose and establish a Philosophical Society. War. Electricity. my first knowledge of it. Partnership with D Hall &c. Dispute in Assembly upon Defence. Project for it. Plain Truth. its Success. 10,000 Men raised and Disciplined. Lotteries . . .'[26] Even this brief extract indicates the breadth of Franklin's political, commercial and social interests. These have led to his universal acclaim, but have eclipsed his contributions to electrical research.

Franklin (1706–90) is one of the United States of America's most famous heroes, the man whose life epitomised three major American aspirations: financial success, moral self-improvement, and social advancement. Hagiographic biographers stress how he rose by his own efforts, pulling himself up out of poverty through his moderation, hard work and honesty. By his early forties, he had made enough money to retire from business, so that he could engage in political affairs, scientific research and philanthropic activities; thirty years later, he was deeply involved in the Declaration of Independence and in drafting the new Constitution.

At the end of the American Revolution, Joseph Banks, President of London's Royal Society, begged Franklin to abandon politics and return to the scientific community, so that he could continue his more important work – electrical research:

> *General Washington has we are told Cincinnatus like return'd to cultivate his garden now the emancipated States have no farther occasion for his sword. How much more pleasant would it be for you to return to your much more interesting more elevated and I will say more useful pursuit of Philosophy . . . Would I could see you abdicate the station of Legislator . . . and return to your Friends here & to those studies which rais'd you formerly to a hight less elevated perhaps but I am sure more satisfactory.*[27]

Until the Second World War, pure science was relatively low on the US national agenda, so that the significance of Franklin's electrical work was neglected. He was celebrated as a founding father who lived out Enlightenment republican ideals – a self-made man of business who became a skilled diplomat, literary writer and philanthropist. Yet as Illustration 1 shows, Franklin was enormously proud of his electrical inventions. Indeed, Banks's plea confirms that Franklin deserved to boast, since he was being congratulated all over Europe not only for his instruments, but also for his theoretical suggestions.

Born in Boston, Franklin was the youngest son among his father's seventeen children. Although his older brothers were apprenticed out, Franklin – originally destined for the Church – was sent to school for a couple of years, but then came home to help in his father's soap and candle trade. Because of his fascination with reading, at the age of twelve Franklin was apprenticed to his brother, a printer, and it was then that he embarked on his life-long programme of rigorous study and self-improvement.

When family relationships deteriorated, Franklin ran away to Philadelphia, where he struggled not only to support himself, but also to purify his character. As he gradually managed to consolidate his financial position, he drew up lists of improving maxims, and compiled check-sheets to monitor his progress. He also took care that his self-imposed frugality and industry were noticed, ostentatiously trundling his printing paper through the streets on a wheelbarrow.

By his early thirties, Franklin had become a prominent Philadelphian running his own successful printing business. He also held the position of postmaster, had founded America's first subscription library, and was publishing *Poor Richard's Almanack*, his popular annual calendar in which he filled any gaps 'with Proverbial Sentences, chiefly such as inculcated Industry and Frugality, as the Means of procuring Wealth and thereby securing Virtue'.[28]

Franklin's own introduction to electricity provides a good example of the ways in which popular journals and travelling lecturers played vital roles in spreading scientific ideas to America as well as throughout Europe. In the 1740s, while still heavily involved in his printing company, Franklin attended a course of natural philosophy given by a visiting Scottish lecturer, whose repertoire included a version of Gray's hanging charity boy. Articles about electrical experiments were being imported from England, and also reprinted in the local press. In 1745, Franklin received an excited letter from an American friend in London. 'As this may I think be very justly stiled an age of wonders', he told Franklin, 'it may not perhaps be disagreeable to just hint them to you . . . now the vertuosi of Europe are taken up in electrical experiments'.[29] Accompanying the letter was a gift: a glass wand for electrical experiments.

Delighted with this new possession, Franklin enthusiastically taught himself how to perform electrical experiments, and later recalled that he had 'much Practice, for my House was continually full for some time, with People who came to see these new Wonders'.[30] Ever the enterprising businessman, he soon converted an unemployed neighbour, Ebenezer Kinnersley, into a successful travelling performer. Kinnersley's advertising posters indicate how entertaining the evenings at Franklin's house must have been, since his lectures included 'Air issuing out of a Bladder set on Fire by a Spark from a Person's Finger, and burning like a Volcano'; 'A curious Machine acting by means of the Electric Fire, and playing Variety of Tunes on eight musical Bells'; 'A Piece of Money drawn out of a Person's Mouth in spite of his Teeth'; and 'A Battery of eleven Guns discharged by a Spark, after it has passed through ten Foot of Water'.[31]

As French and Spanish invaders threatened Philadelphia, Franklin became embroiled in political activity. Nevertheless, he launched himself into an intensive programme of electrical investigations, sending a series of letters to England describing his results and outlining his novel theoretical explanations. These letters circulated among London friends, and reached wider audiences by appearing in the *Gentleman's Magazine*. The journal's editor, another Enlightenment entrepreneur, seized the opportunity to earn a substantial profit by publishing some of them as a small book, which was soon translated into French.

Franklin's distance from Europe had resulted in his inadvertently reinventing some existing devices, but it also gave him the intellectual liberty to develop his own ideas. The originator of new instruments and theories, Franklin rapidly became one of Europe's most influential authorities on electricity. It was in France rather than England that he first shot to fame, becoming involved in the rivalry between Nollet and other eminent French natural philosophers. As part of their manoeuvring, the king was invited to witness some of Franklin's electrical parlour tricks (the magical portrait was, of course, diplomatically included in this performance). Encouraged by the royal enthusiasm, Nollet's competitors decided to stage a dramatic trial.

Many people thought that lightning might well be electrical, but it was Franklin who had proposed a viable way to test this hypothesis. He suggested making a kind of sentry-box with a long rod pointing up from the roof to 'draw off' electricity from thunder clouds overhead. Inside, the sentry-experimenter would stand on an insulated stool, protected from the rain as well as the lightning. In 1752, following this idea, two French researchers set up a tall insulated pole in the small village of Marly (Franklin, so keen on promoting temperance, would perhaps not have approved of the three wine bottles used to help support the apparatus). Bored by waiting for thunder, the experimenters had already returned to Paris by the time a storm appeared. But two local assistants – a retired soldier and the local priest – valiantly drew sparks by touching the pole with a brass wire mounted in a glass handle.

This simple yet risky demonstration successfully identified natural lightning with man-made electricity. It established Franklin's reputation not so much because of its theoretical results, but rather because of its practical implications, since it confirmed Franklin's claim that grounded rods could protect high buildings from being burnt down during a storm. Following the success at Marly, other investigators enthusiastically tried to replicate the results, rendered deceptively straightforward by diagrams such as the one near the top left of Illustration 8.

Franklin himself prudently never carried out some

of his own proposed experiments, which resulted in at least one fatality when adventurous experimenters clambered about on rooftops trying to record a charge. In England, natural philosophers competed to be the first to gather lightning. Benjamin Wilson suddenly abandoned his stage performance as Henry IV when a thunderstorm started. Racing out in his royal robes to the bowling green, he stuck a curtain rod in a bottle to collect the electric fluid. Ironically, Wilson – who was a society artist as well as one of the Royal Society's electrical experts – later became one of Franklin's most outspoken opponents.

However, although the successful trial at Marly was then widely interpreted as a crucial experiment, it has largely been forgotten. Instead, because Franklin has become a founding father of electricity as well as of the United States, it is his own slightly later version that is celebrated. Over in Philadelphia, several weeks away from the European news, Franklin made a strong kite from a silk handkerchief and a cedar cross carrying a sharp, pointed wire. To the kite's string, he attached a metal key carefully wrapped in dry silk to insulate it, sensibly stationing himself under cover. By flying the kite into a thunder cloud, he caused sparks to stream from the key and successfully charged up a Leyden jar; '& thereby', his own account triumphantly concludes, 'the Sameness of the Electric Matter with that of Lightning compleatly demonstrated'.[32]

Franklin's kite experiment may not have increased the knowledge gained during the Marly test, but it has entered scientific mythology because of its symbolic significance. Franklin's kite promotes a triumphal version of science in which progress is made through individual momentous discoveries. Receiving the accolades formerly reserved for military heroes or religious leaders, scientists have been converted into the secular saints of modern society.

Franklin became the modern incarnation of Prometheus, the Greek hero who aroused Zeus's fury by stealing fire from heaven – the spark of divine wisdom – to give human beings the power of reason. Prometheus also represented freedom from political oppression, and the jagged streak of lightning that frequently appears in pictures of Franklin represents his political ideals as well as his electrical inventions. The innovatory electrician was also a revolutionary politician who became idolised in France for 'snatching lightning from the sky and the sceptre from the tyrant'.[33]

# 2 Knobs or Points?

Nollet, fumed Franklin, 'speaks as if he thought it presumptious in man to propose guarding himself against the *Thunders of Heaven*! Surely the Thunder of Heaven is no more supernatural than the Rain, Hail or Sunshine of Heaven, against the inconvenience of which we guard by Roofs & Shades without Scruple.'[34] Like Nollet, many people regarded lightning as a sign of God's displeasure at a sinful world. Churches seemed to be particularly vulnerable, and a special prayer was recited during religious ceremonies to consecrate the bells, which were often rung loudly as an antidote to storms. For natural philosophers, demolishing such traditional beliefs provided a good opportunity to demonstrate the superiority of rational analysis. Phrases such as 'enlightening the gullible masses' abound in eighteenth-century writing, and even the egalitarian Franklin referred to 'the superstitious prejudices of the populace'.[35]

Sneering at the rituals of 'the Romish church', Franklin preached the scientific explanation. Since, he argued, church steeples were the highest structures in most towns, electricity would be conducted from the cross down through other metal components, thus making churches particularly vulnerable. He suggested installing a pointed rod that, connected to the ground by wires, would conduct the lightning away from the building to be harmlessly discharged into the earth. By introducing lightning rods to protect churches, large buildings and ships, Franklin not only saved money and preserved lives, but also helped to transfer authority from religious institutions to scientific ones.

Using texts, pictures and instruments, Franklin and his allies actively promoted his new invention. One enthusiast was concerned about the risks encountered by women wearing metal hairpins and silk shoes, and recommended – apparently in all seriousness – that a lady should equip herself 'with a small chain or wire, to be hooked on at pleasure during thunderstorms . . . from her cap . . . down to the ground'.[36] The cityscape in Franklin's portrait (upper right of Illustration 1) compares a solid brick building carrying its prominent rod with a lightning-devastated scene of destruction, as a church topples and a house explodes. This was an artificial view picturing educational wooden models that were specially designed to advertise the virtues of electrical protection. Beautifully crafted out of mahogany, these miniature collapsible steeples and 'thunder houses' simultaneously amused and instructed audiences when they were detonated with small amounts of gunpowder.

Within a few years, American houses, churches and ships were routinely protected by Franklin's rods, but there was considerable resistance in Europe. Although the papal authorities recommended safeguarding churches with lightning rods, many local parishioners and clergymen opposed taking what they viewed as sacrilegious measures against God's will. In addition to this religious reluctance, natural philosophers engaged in extensive theoretical debates about the shape and location of the rods. For Franklin, it was essential that the rods be tall and pointed, since the effectiveness of sharp points in '*drawing off* and *throwing off* the electrical fire' was central to his theories.[37] But opponents such as Nollet and Wilson recommended smaller, rounded rods, arguing that it was foolish to construct devices explicitly designed to attract electricity rather than avoid it, particularly since electricity was evidence of divine wrath.

The controversy gained urgency as people continued to die and prominent buildings were damaged. Contemporary disasters included St Bride's church in

London, a gunpowder store in Jamaica and a cardinal's palace in Italy. Matters came to a head in 1777 when the British army's arsenal at Purfleet was damaged, even though it was protected by pointed rods. Franklin's critic Wilson insisted that lightning conductors needed to be topped with rounded copper knobs, and he decided to stage a trial in the Pantheon, an elegant London concert hall.

Illustration 9 shows Wilson's two giant horizontal cylinders, attached to several hundred yards of wire. They are being charged up by an electrical machine to create an artificial cloud hovering over an accurate scale model of the arsenal (on the right). Able to move along rails in order to simulate the relative motion of clouds and buildings, this replica building could be fitted with either blunt or pointed rods at various heights, and could be tested either dry or wet. The central pronged device helped to analyse the effects of neighbouring clouds, while the man on the left is shown holding flaming material that has been set alight electrically.

It is tempting to laugh at this contest between rounded and sharp rods: it seems like a real-life equivalent of the argument in *Gulliver's Travels* about which end of an egg to eat first. However, this debate resonated throughout English society for many years because it was about far more than

*Phileo Trans. Vol LXVIII. Tab.III. p. 245*

*A View of the Apparatus and part of the Great Cylinder in the Pantheon.*

*Illustration 9*: Benjamin Wilson's experiment at the Pantheon in 1777. *Philosophical Transactions* 68, 1778, facing p. 244. (Cambridge University Library)

knobs and points. Political interests permeated the supposedly objective experiment. The American Franklin, Wilson emphasised, should no longer be regarded as an Englishman because he 'was becoming one of the Chiefs of the Revolution'.[38] Moreover, he frequented the Club of Honest Whigs, centre for enlightened reformers, and associated with political and religious dissenters.

Wilson, a Tory with a wealthy artistic clientele and considerable influence at court, persuaded King

George III to witness the dramatic test in person. The king was totally won over, and Wilson's knobs immediately replaced Franklin's points at Buckingham-House. The President of the Royal Society, who was a staunch Franklin supporter, resigned over the issue, allegedly rebuking the king that his 'prerogatives . . . do not extend to altering the laws of nature'.[39] Enlightenment wits delighted in making fun of the episode:

> *While you, great George, for safety hunt*
> *And sharp conductors change for blunt,*
>   *The nation's out of joint:*
> *Franklin a wiser course pursues,*
> *And all your thunder fearless views,*
>   *By keeping to the point.*[40]

This epigram was far from being a trivial piece of humour. Its contemporary bite depended on the close ties between politics and electricity, a relationship that contravened Enlightenment ideals of gaining neutral knowledge about the natural world.

# 3   The Business of Medicine

Natural philosophers were convinced that electricity, weather and life were intimately linked to one another. Citing Franklin's experiments, which had confirmed that lightning was the same as artificial electricity, they argued that atmospheric electricity was involved in other alarming events, such as earthquakes and volcanoes. Franklin and his contemporaries were also fascinated by the northern lights (*aurora borealis*), whose displays were exceptionally spectacular in the early eighteenth century, often being visible as far south as England.

Diarists were meticulously recording weather patterns and attempting to correlate them with their own bodily symptoms and psychological states. Thunderstorms had long been associated with lassitude and frayed nerves, while a large surge in the number of births at Lyons was attributed to that year's exceptionally strong north wind, with its high electrical content. Enlightenment electrical investigators tried to establish more systematic links between vitality and electricity, which was often depicted as the great 'vivifying principle' of the earth and its living inhabitants. As Rackstrow explained (albeit rather vaguely): 'This Ætherial or Electrical Fire I take to be the vivifying spirit that resides in Air, which, taken into the lungs of an animal . . . swiftly passes over the whole body.'[41]

Since the wealth of most countries lay in their agricultural resources, attention soon focused on the possibilities of electrifying plants. Installing elaborate devices in greenhouses, experimenters carefully regulated conditions and set up controlled tests in order to assess claims that electricity stimulated plant growth. Despite initial optimistic predictions of increased yields, by the end of the century this line of research had largely been abandoned.

In contrast, the accidental shocks endured by early investigators had demonstrated beyond any doubt that electricity strongly affected human bodies. After more systematic attempts at self-experimentation, many electricians – including Franklin – decided to examine the reactions of sick people. Following exciting reports from Geneva that a locksmith's paralysed arm had been restored to normality, within a few years electrical medicine had taken off all over Europe.

One of the earliest English converts was the Methodist preacher John Wesley, who learnt about electricity from Franklin's published letters and later bought his own machine. Initially setting aside an hour a day to treat his flock, within three years he was overwhelmed with patients and completing his own book on electrical medicine. Collating case-histories from all over England, Wesley listed illnesses – many 'of the nervous kind' – that ranged alphabetically from 'Agues, St Anthony's fire, Blood extravasated, Bronchocele, Coldness in the Feet' to 'Toe hurt, Tooth-ach, Wen'.[42]

As these examples suggest, eighteenth-century people classified complaints differently from us. As well as using unfamiliar labels, they concentrated on the discomfort experienced by each individual rather than on the symptoms that characterise a general disease. Wesley and his fellow electrical physicians paid close attention to details of behaviour, social circumstances and other indicators that seem extraneous to many modern doctors. One reason for this was that most patients were paying for their treatment and so demanded a high level of personal care, which was, of course, often in itself thera-peutically beneficial. Wesley himself may have been motivated by Christian charity, but most medical practitioners were competing for their clients' money. Enlightenment medicine was big business in England, especially in London, where profits could be extremely high for a doctor who managed to gain his patients' trust.

Since many illnesses defied the knowledge and treatments available, becoming a successful medical man did not depend solely on expertise, but required developing a good bedside manner, making influential contacts, and deploying resourceful self-marketing strategies. Medicine was highly competitive, and society physicians often denigrated their rivals as quacks or charlatans, as in Illustration 10. In the 1780s, London's most famous electrical physician was James Graham, shown on the left in this caricature, standing on a gambling table and accompanied by a quacking duck, a typical Enlightenment visual pun. Perched precariously on two insulating stools in front of his Leyden jars, Graham straddles his giant 'Prime Conductor', which points at his major medical rival, the German Gustavus Katterfelto.

Flanked by his Leyden jar, thunder house and miniature aurora, Katterfelto retaliates with his own 'Positivley Charg'd' conductor linked to a firing cannon, while sparks flash from his fingertips. The self-styled 'Doctor' Katterfelto had leapt to fame by dispensing patent medicines during an influenza epidemic. For a couple of years he had attracted huge London audiences to his experimental lecture demonstrations, cleverly exploiting the magical

*Illustration 10*: 'The Quacks': James Graham and Gustavus Katterfelto, 1783 engraving. (Wellcome Institute)

appeal of electrical machines, projecting microscopes and other equipment. Later, Katterfelto's reputation plummeted, and he was forced to trail around provincial lecture halls to eke out an existence.

Graham's career was more dazzling. A Scot who had practised electrical medicine in America, he opened a grandiose Temple of Health in central London that for a while drew over two hundred patients a day. Promising to restore 'that full-toned juvenile Virility which Speaks so cordially and Effectually home to the Female Heart', Graham constructed his major selling point, his notorious

Celestial Bed. Surmounted by an elaborately decorated canopy lined with mirrors, this enormous tiltable edifice was set in a private room through which flowed soft music, spicy aromas and circulating electrical fluids. From two giant Leyden jars in a mahogany case, brass rods conducted restorative electricity through a fiery dragon. White-robed Vestal Virgins – including Emma Hamilton, later Lord Nelson's lover – assisted couples to enjoy the reinvigorating powers of this electric storm, perhaps encouraging them to admire the electrical sparks reflected in the chandeliers. Those unable to pay £50 could indulge in less splendid electric treatments, or benefit from Graham's stimulating lectures on sex and electricity. Forced to diversify after the Temple's financial collapse, Graham moved into mud baths (which also proved an unprofitable venture).

Although these two philosophical rivals are here denigrated as quacks, there were no hard boundaries either between Katterfelto and more didactic lecturers such as Walker and Priestley, or between Graham and eminent society physicians coming from a more conventional background. Modern medicine is so well established that patients have only two choices: orthodox or alternative. But in the eighteenth century, medical practitioners ranged along a continuous spectrum which ran from entrepreneurs like Graham to the prestigious university-educated members of the College of Physicians. The power to choose between competing therapies lay not in research laboratories and government funding, but in the purses of wealthy patients.

Towards the end of the century, electrical therapists had accumulated enough reports of successful cures to convince doctors as well as sick people that their treatments were worth trying, particularly when all other attempts had failed. As part of their own attempts to become recognised as valuable contributors to society, some innovative natural philosophers and physicians explored the claims of electrical medicine and tried to set it on a more solid basis. As they refined their methods, they started to publish textbooks describing sophisticated equipment. They gave detailed instructions for treating particular complaints, and discussed failures as well as successes. Illustration 11, showing a small girl being treated electrically, forms the frontispiece of a long book on electricity. This was no quack publication. It was written by George Adams, one of London's most prestigious instrument makers, who carefully explained the labelled diagram in technical terms. The tranquil expressions of the girl and her mother, reinforced by the practitioner's calm

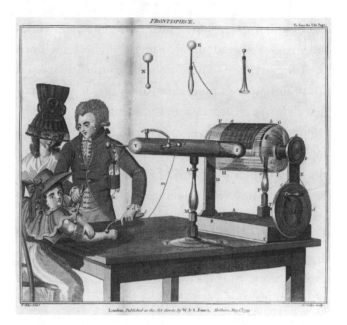

*Illustration 11:* Monitoring electric therapy with Timothy Lane's electrometer. Frontispiece of George Adams's *An Essay on Electricity, Explaining the Theory and Practice of that Useful Science; and the Mode of Applying it to Medical Purposes*, London, 1799. (Cambridge University Library)

confidence, reassure hesitant readers that this is a safe procedure of genuine value.

Enlightenment rationalists worked hard to push a wedge between enterprising physicians who were sanctioned by society, and their neighbours on the medical spectrum, those innovators who could be dismissed as charlatans. To marginalise competitors like Graham, Adams and his entrepreneurial colleagues adopted diverse tactics. These included satirising rivals as quacks, and writing their own promotional sales literature (often making exaggerated claims) for their new instruments and treatments.

Physicians also developed techniques to discriminate between different therapies. The method of blind testing, now routine for new drugs, was first designed in the 1790s to refute the advertising claims being made for metal 'electric tractors'. Resembling tuning forks, these brass and iron tractors would supposedly 'draw out' disease electrically by being stroked along the sick person's body. Dividing their patients into two groups, doctors at Chester Hospital demonstrated that there was no difference between the effects produced by the genuine article and by wooden imitations.

Although the benefits of electrical medicine were not universally accepted, it did attract respected adherents, including Franklin. To us, eighteenth-century electrical treatments may seem laughable, or even fraudulent, yet many therapists fell on the legitimated side of a boundary that was becoming increasingly solid.

# 4 Therapeutic Shocks

Based in Paris, Nollet regarded himself as Europe's leading electrical expert in the middle of the century. As soon as he heard about the successful cures of paralytics being carried out in Geneva, he anxiously set about establishing his own authority as a therapist, and obtained special permission from the war minister to convert a room in the Invalides hospital into a temporary research centre. When he failed to achieve convincing results, he promptly exonerated his own techniques by blaming other factors: in particular, he insisted, one could not expect the bodies of Swiss working-class men to respond in the same way as those of élite French natural philosophers.

The situation worsened for him when separate groups of Italian investigators reported dramatic progress with novel electrical techniques. These included using glass tubes filled with different substances, such as opium and perfumes, and electrifying people who were holding powerful purgatives in their hands. In 1749, Nollet decided to tour Italy and witness these miraculous remedies for himself. Protesting disingenuously that it was in his own interests to find positive evidence for electricity's medical virtues, he was in reality a far from neutral inspector, but was a prejudiced observer who sneered: 'I really fear that I have nothing to purge here except imagination.'[43]

As Nollet's remark indicates, purging was a major preoccupation of eighteenth-century gentlemen, and his first visit in Turin was to the consulting rooms of a purgation specialist. Nollet subjected himself to a half-hour electrical session (briefly interrupted when the cord on the machine's wheel got tangled). Electrification caused sparks to fly from his forehead, but – or so he reported – was otherwise ineffective. But was it? Nollet did admit to some symptoms, but glibly attributed them to a large glass of iced lemonade and his chronically sensitive digestive system. Similarly, it was easy for him to dismiss servants' testimonies of success by questioning their honesty, accusing them of accepting bribes, exaggerating to impress their friends, or of secretly taking oral laxatives. Adopting similar rhetorical tactics in Venice, Naples and other Italian cities, Nollet sought to restore his threatened reputation.

Looking back, it seems clear that self-examination was simply not a reliable way of ascertaining whether electrical therapies worked. Nevertheless, despite Nollet's damning report of his Italian experiences, several natural philosophers and physicians remained sufficiently impressed by the stories they were hearing to try out electrical treatment themselves. Paralysed patients, presumably grasping at any hope, travelled long distances to visit Franklin. Three times a day, he gave them strong shocks from six-gallon Leyden jars, but he regretted only managing to provide them with temporary relief.

In England, as in Italy, many of the early trials seem to have been carried out on servants and farm workers, who were in no position to object to becoming experimental subjects. One typical case from 1756 was 'Elizabeth Stokes, aged twenty-three, a very lusty and healthy woman, [who] was, in the beginning of January last, seized with a rheumatic kind of pain in her right arm, particularly about the wrist; and in two or three days time afterwards, the finger and thumb of that hand contracted up so close, that they could not be opened with any force the girl herself could use to them.'

Stokes was taken to Shrewsbury Hospital, where her hand was electrified until sparks flew, scrubbed vigorously with a 'flesh-brush', and bound in flannel. After a week of this treatment, 'the shock began to be so painful, that she desired to be excused from it any more'. Sent back to her employers, she suffered a relapse after ten days of washing clothes continuously, and was subjected to even more intensive shock therapy. Fitted with a permanent drain in her arm to let blood, she was kept as an in-patient for several weeks before being discharged. From the medical report, it is impossible to tell whether she was completely cured – as her doctor proudly claimed – or whether continuing with her domestic chores (however heavy) seemed preferable to receiving repeated shocks.[44]

Throughout the 1750s and 60s, numerous small tracts on electrical medicine were published, and reports from all over Europe frequently appeared in the *Philosophical Transactions* and other journals. Electricity was apparently being used successfully to treat all sorts of paralytic complaints, as well as an impressive range of other ailments: abscesses, con-vulsions, blindness, rheumatism, fevers, swollen joints – to say nothing of the ubiquitous gout and constipation. Many of these accounts included affidavits from grateful patients, although only gentlemen wrote in their own voice. This was not just an issue of literacy: élitist medical writers did not believe that working-class people could be trusted as dependable observers. So although these case-studies gained authenticity by giving the subject's name, address and occupation, they had to be countersigned by a reliable witness – such as a clergyman or a doctor – to vouch for their truthfulness.

Yet despite the successes, many critics doubted the efficacy of this new treatment. The machines were cumbersome to use, and were annoyingly temperamental in their behaviour, especially in damp weather. While electrical therapy had scored some dramatic and well-attested cures of paralysis, other paralysed patients remained totally unaffected. Analysing this conundrum from our perspective, we would say that similar symptoms can be manifestations of different underlying diseases. Many therapists were troubled by not knowing the size of shock to prescribe. Perhaps, Franklin wondered, it might be better to give a couple of hundred small shocks over a longer period of time than a few massive ones, as he had done. This problem was compounded by being unsure how to compare the output of different machines.

To tackle these difficulties, instrument makers started to market new instruments that would measure the charge being delivered. Illustration 11 (on page 89) shows Timothy Lane's electrometer – the small adjustable attachments and wires linking the prime conductor to the Leyden jar. Using specially shaped directors with glass handles, the operator can ensure that an appropriate shock is applied in exactly the right place to his young patient. She sits on an insulated chair, her afflicted arm carefully wrapped in flannel. In modern terms, the sudden drop in potential as the battery of Leyden jars discharged could reach several thousand volts. Devices like Lane's enabled therapists to use far less drastic techniques. By the end of the century, the emphasis lay on gentleness, gradually building up the shock from a very low level until it became effective. This softer, more persuasive approach must have helped to increase the number of paying patients.

By the 1780s, several of Britain's major hospitals had installed their own electrical machines, and some eminent doctors regularly referred their more intractable cases to the electrical therapist. At St Thomas's Hospital in London, John Birch ran the electrical department for over twenty years, surviving the scepticism of some of his more conservative colleagues. Far from being a quack, Birch was a respected surgeon whose private patients included members of the royal family. As advocates successfully marketed their improved techniques, demand grew sufficiently to justify opening specialised institutions. The London Electric Dispensary, which opened in the City Road in 1793, thrived for at least twenty years, treating about 300 patients a year and claiming to cure half of them.

Compared with England, medicine in France was far less determined by patient choice because the state exerted a stronger influence. This contrast affected many areas, and is one of the major reasons why the patterns of science's history are so different in the two countries. Whereas English experimenters were obliged to finance themselves, public money was available to fund research in France on a wide range of topics both before and after the Revolution of 1789. For example, scientific academies in Paris and the provinces often set prize essay competitions, and some of these stimulated further investigation into electrical medicine. On the other hand, by refusing money or a license to practice, the state could also control the type and extent of experimental work being carried out.

In 1776, the newly founded Royal Society of Medicine in Paris decided to put its weight behind electrical therapy. According to them, the links between health, weather and electricity were clear cut: 'Men and animals are generally more vigorous in the winter and at all times when the air is dry; heat and humidity enfeeble them. Cold and dryness render electricity stronger; heat and humidity diminish it.'[45] Since all that remained – it appeared –

was to set this straightforward relationship on a quantitative footing, the Society solicited weather reports from doctors all over France, instructing them how to obtain accurate measurements of cloud cover, atmospheric electrical charge, temperature and so on.

Although this ambitious national project yielded no immediately conclusive results, the Society continued to investigate therapists' claims. Just as Nollet and the English physicians had been suspicious of evidence from the wrong class of people, so too, official committees sanctioned only certain practitioners. Thus an electrical performer with the stage name of 'Comus' was granted a medical licence and given clinical facilities to pursue his treatment of epilepsy. On the other hand, some campaigners were hampered by not moving in the right circles. One of these experts was Jean-Paul Marat. Despite the cogency of his arguments against electrical medicine, this scientific outsider was ignored – until, that is, he was killed in a bath for his political beliefs.

# Part IV
# Sparks of Imagination:
# Theories

*The greatest part of natural philosophers acknowledge their ignorance ... There is no room to doubt that we must look for the source of all the phenomena of electricity only in a certain fluid and subtile matter; but we have no need to go to the regions of imagination in quest of it.*

Leonhardt Euler, *Letters to a German Princess, on Different Subjects in Physics and Philosophy*, 1795

Robert Symmer became known in France as the '*philosophe déchaussé*', the 'barefoot philosopher'. A Fellow of the Royal Society and a financial administrator for King George III, Symmer was intrigued by the crackling sparks emitted from his stockings when he got undressed in the dark. To keep warm, he had developed the habit of wearing white silk stockings over his black wool ones during the winter mornings, and swapping them round at lunch time. One day in 1758, as he stripped off this double layer, the two different stockings obstinately clung together; when he forcibly pulled them apart, they swelled up as if they still contained a phantom limb. Stuck to the wall, they would eerily dance in the breeze and attract small particles a metre or more away. By using four silk stockings to charge up a Leyden jar, reported Symmer, he 'kindled spirits of wine in a tea-spoon . . . [and] felt the blow from my elbows to my breast'.[46]

Symmer wrote meticulous accounts of his experiments, carefully recording details of the weather and the weights needed to tear his stockings apart. In spite of this thoroughness, many electrical experts regarded his experiments with either mirth or indifference. But in Paris, Nollet seized on Symmer's conclusions as ammunition in his protracted campaign against Franklin. Symmer's stockings vindicated Nollet's belief that there were two types of electric fluid, whereas Franklin insisted that there was only one. Europe's natural philosophers were divided into two camps, the one-fluid and two-fluid supporters. Neither side had completely convincing evidence to support its case, but immediately interpreted any new results to lend weight to its own arguments.

It is easy to laugh at the affair of Symmer's socks. It seems incongruous that his idiosyncratic dressing habits should provide the basis for a serious scientific controversy, yet the development of Enlightenment electrical theories was clouded by confusion and conflict. Enlightenment philosophers often declared that natural knowledge was steadily progressing, but their high-sounding claims concealed a more muddled reality. For instance, the Leyden jar patently worked, yet contravened existing theoretical models. Theories based on interacting particles explained some electrical effects, but failed to cope with others. Fluid models generally seemed to work quite well, partly because they could be expressed vaguely. But how many fluids? And what were they made of?

Franklin played a crucial role in the debates about electricity's behaviour, introducing a new one-fluid theory in the middle of the century. Yet even his version left some observations unexplained. Eighteenth-century electrical theories are often hard to understand because they use poorly defined terms (such as 'atmosphere' and 'fluid') and gloss over inconvenient results that don't fit in. Especially in England, strange fluids called 'aethers' were often invoked to fulfil religious rather than scientific

requirements. Towards the end of the century, continental researchers used high-precision experiments
and mathematical techniques to approach electricity
in a completely new way, which would form the basis
for later research in the nineteenth century.

# 1 Problems

One outstanding puzzle was the relationship between electricity and magnetism. Nowadays, even people with little scientific education regard them as being firmly bonded together. To account for their behaviour, we use terms like 'field', 'energy' and 'electromagnetic force', even without necessarily being completely sure what they mean. Yet these are all nineteenth-century innovations. Understanding electrical theories during the eighteenth century entails divesting ourselves of all these later notions. Instead, we need to appreciate the conceptual basis of these early electrical investigators who were struggling to find explanations for the strange phenomena that their inventions produced.

Unlike later scientists, Enlightenment theoreticians regarded electricity and magnetism as separate powers of nature. Although it was well known that thunderstorms could ruin ships' compasses, most natural philosophers remained convinced that the two phenomena were different. They drew up comparative tables. Electricity was produced artificially, affected a wide range of substances – including people – and could be heard, seen and smelt. Magnetism, on the other hand, occurred naturally, apparently passed unchanged through anything not made of iron or steel, and was undetectable by human beings.

Electricity and magnetism were also studied differently. Because compasses were so important for navigation, practical expertise was rooted in seafaring communities. As natural philosophers took over sailors' knowledge and techniques, they concentrated on mapping the patterns of the earth's magnetism and constructing more accurate measuring instruments. In a sense, they regarded the whole world as their laboratory. In contrast with these traditional and international activities, electrical practitioners were dealing with a new phenomenon, one that had been generated by their own colleagues. They operated in local environments – private studies, lecture halls and theatres – developing exciting performances that would convince public audiences of their control over nature.

To back up their claims of expertise, experimenters needed to provide believable explanations of electricity's perplexing behaviour. This proved hard to achieve, and by the end of the eighteenth century there was still no consensus. Much as

Enlightenment natural philosophers advertised their rational approach, they often resorted to postulating substances that exhibited the most peculiar characteristics. Another tactic was to accept a new theory because it satisfactorily explained a new instrument or experiment – even though the theory failed to cope with other observations, it could, it was hoped, be patched up later.

The dilemmas confronting Francis Hauksbee, one of the first investigators, provide a good example. Like many experimenters, he visualised electricity as a cloud of tiny corpuscles, which he called 'effluvia'. Hauksbee's effluvia had very strange properties. For one thing, they were 'subtle', by which eighteenth-century writers meant that they were invisible and weightless, and could pass through solid substances like glass without trace. Yet at the same time, they were obstructed by flimsy cloth, generated light by friction, and could make cotton threads stand up stiffly. Moreover, they were all electrically attractive, which made it hard for Hauksbee to explain how charged objects such as gold leaves could fly apart from one another.

Just as electricity confounded Enlightenment experimenters, so too, their convoluted prose must surely have confused their readers. This is a sample of Hauksbee's writing, in which electricity slips almost imperceptibly from being a swarm of effluvia to become a singular substance that behaves like a solid:

> [This experiment] *shews the subtilty of the Effluvia, the Body from which it is produc'd seeming to be no Impediment to its motion; Besides it seems very much to resemble or emulate a Solid, since Motion may be given to a Body, by pushing the Effluvia at some distance from it: But what is still more strange is, That this Body (I presume to call it so) altho' so subtil as seemingly to perviate Glass, will not ... affect a light Body thro' a piece of Muslin: Now whether the Muslin absorbs the Effluvium, or what other Laws it may be subject to, I cannot tell, but sure I am 'tis very amazing.*[47]

Hauksbee's last gasp of amazement does imply defeat. Although he and his successors continued their attempts to pin down electricity's elusive nature, they remained unable to provide one comprehensive theory that would single-handedly account for all of their experimental observations.

During the seventeenth century, many natural philosophers had become increasingly determined to find mechanical explanations for physical phenomena. Eager to distinguish themselves from astrologers and alchemists, they condemned attempts to account for the mysterious effects of attraction and repulsion by hidden powers or occult sympathies. Instead, they preferred to envisage streams of tiny invisible particles that acted on matter by direct contact. Some of them gave their particles distinctive properties – such as stickiness or fattiness. Nevertheless, despite differing in details, they shared an overriding concern to provide mechanical models for electrical activity.

The leading proponent of this type of view was the French natural philosopher René Descartes, who suggested that the universe is packed with swirling vortices of corpuscles that God had created in different shapes and sizes. The central showpiece of Descartes' complex edifice was magnetism, due – according to him – to two types of particles, each rushing in opposite directions through the internal pores of magnetic materials. Using these counter-flowing currents, Descartes explained repulsion by visualising magnets being pushed apart as swarms of corpuscles competed for space; conversely, attraction resulted from the magnets being driven towards one another from behind. For electricity, Descartes suggested a similar mechanism which involved the behaviour of particles in pores of different shapes, although he was less precise about the specific details. Convoluted though Descartes' arguments may seem, they were extremely influential, and throughout the eighteenth century many natural philosophers treated electricity as a corpuscular fluid.

From around 1700, most English experimenters were, as a matter of principle, strongly opposed to Descartes. Enlightenment men of science often claimed that they were engaged in a collaborative, international project to uncover the truths of nature. In practice, natural philosophy was riven with national interests: the intellectual debates between the rival supporters of men like Descartes, Newton and Leibniz were strongly coloured by chauvinist loyalties. England was at war with France for much of the century, and patriotic Newtonians waged intellectual hostilities against Descartes, exultantly announcing 'the routing of this Army of Goths and Vandals in the Philosophical World'.[48] Reciprocally, French resistance towards Newtonianism was fuelled by Jesuit educators on religious grounds, and it was not until the second half of the eighteenth century that most continental natural philosophers adhered to some modified version of Newton's ideas.

The emergence of a Newtonian world view is seen as one of the triumphs of Enlightenment rationality, but Newton's own works had been riddled with internal contradictions and inaccuracies. By the end of the century, although most natural philosophers described themselves as Newtonians, they in fact embraced a wide variety of views. Ideas consistent with a Cartesian framework (that is, one based on Descartes' ideas) permeated accounts of electricity. One outstanding example is the famous Swiss mathematician Leonhardt Euler, acclaimed by his contemporaries as a great genius, who is often – misleadingly – said to have converted Berlin's Academy of Sciences to Newtonianism.

Euler published a best-selling book that, promptly translated into several languages, spread his ideas about natural philosophy throughout Europe. Adopting the literary device of softening technicalities by writing letters addressed to a German princess, Euler covered a wide range of topics, including electricity. This supposed Newtonian presented a very Cartesian account that would, he promised his princess, provide her with a clear road through the maze of mysterious electrical phenomena. Unfortunately, his explanations are not immediately illuminating:

> ... [B]odies evidently become electric, only so far as the elasticity, or the state of the compression in the pores of bodies, is not the

*same as every where else ... the prodigious elasticity with which the ether is endowed, makes violent efforts to recover its equilibrium, and to restore every where the same degree of elasticity, as far as the nature of the pores, which, in different bodies, are more or less open, will permit; and it is the return to equilibrium which always produces the phenomena of electricity.*[49]

Like the earlier lengthy quotation from Hauksbee, Euler's prose – taken from an educational textbook! – illustrates the opacity of eighteenth-century electrical explanations.

# 2    Fluids and Atmospheres

During the 1730s, Cartesian France became the major location for theoretical work. In England, the pioneering investigations carried out by men like Hauksbee and Gray had demonstrated the astounding effects of electricity. Yet as the President of the Royal Society wrote plaintively to a French colleague, 'no one here is offering to explain them'.[50] In Paris, Nollet was serving his electrical apprenticeship with Charles Dufay, who managed the Royal Botanical Garden and belonged to the Academy of Sciences. Dufay significantly affected electrical research because he approached it systematically, insisting on methodical processes of investigation and reporting.

Above all, Dufay was determined to resolve the question of repulsion. He became convinced of its reality after a simple experiment that entailed chasing a gold leaf around his room with an electrified glass tube. He found that the leaf was attracted when unelectrified, yet repelled after electrification by the tube. Pursuing this line of inquiry, he concluded that there are two types of electricity: 'resinous' electricity, produced by rubbing substances like amber and wax; and 'vitreous' electricity, obtained from glass or crystal. As he explained in a letter to London's Royal Society: 'The Characteristick of these two Electricities is, that a Body of the *vitreous Electricity*, for Example, repels all such as are of the same Electricity; and on the contrary, attracts those of the *resinous Electricity*.'[51]

But how did electricity affect people? To answer this, Dufay converted his own body into an experimental object by having himself suspended like one of Gray's charity boys (Illustration 5, on page 45). As a trained observer, he claimed he could be trusted to provide objective evidence and report faithfully on the pain and other bodily experiences provoked by different electrical contacts. To explain his results, Dufay relied on Cartesian models of swirling vortices. People and objects are, he surmised, surrounded by 'atmospheres' that greedily pull electrical matter into themselves, an exchange that manifests itself by noisy sparks.

Loyal Cartesians welcomed Dufay's electrical conclusions because they offered tangible evidence that vortices were no mere conjectures, but really existed. Unfortunately, these rotating whirlpools could not explain the simple observation that electrified objects are repelled or attracted along straight lines. To resolve this problem, Nollet – who succeeded Dufay after his teacher's early death – imagined that all electrical effects are due to a single type of electrical matter, whose behaviour depends on whether it is leaving a body in a diverging jet or entering it in a converging ray (upper part of Illustration 7, on page 53). This version of Dufay's resinous and vitreous electricities enabled him to account for repulsion and other puzzling phenomena. Nollet's illustration of Symmer's stockings shows them temporarily inflated by small internal spurts of electrical fluid; he attributed their subsequent collapse to interference between resinous and vitreous jets.

Despite the confidence displayed by Illustration 7, Nollet's theory failed to explain the Leyden jar satisfactorily. Even so, around the middle of the century Nollet was acclaimed as Europe's leading electrician. His supremacy was challenged by Franklin, an American outsider who took a totally new approach. Steeped in Newtonian thought, Franklin was largely ignorant of the Cartesian-based electrical theories. Unlike Dufay and Nollet's flowing fluids, Franklin's stationary atmospheres of electrical matter clouded around electrified objects and pushed them apart (in Illustration 8, on page 59, see Figures 15 and 18).

In contrast with the prevailing models of two types of electricity, Franklin declared that there was only one. Perhaps reflecting his business experience, Franklin's fluid behaved rather like money: just as a bank account may be in credit or overdrawn, an object may have more or less than its normal equilibrium amount of electricity, thus making it appear to be positively or negatively charged. Franklin replaced terms such as 'vitreous' and 'resinous' with 'plus' and 'minus'. Rubbing glass, he maintained, does not produce electricity but either adds it or takes it away.

Franklin's major coup was to account for the Leyden jar more coherently than his rivals. Through his experiments, he determined that the electric charge is held inside the glass itself, a result that he confirmed by using glass plates instead of the bottle. On his account, it is glass's special impermeability to electric fluid that enables the jar to function. The positive charge on the jar's outer coating is exactly equal to the opposite negative charge inside it. 'There is really no more the total quantity of electrical fire in the phial after what is called its *charging*, than before, nor less after its *discharging*', he wrote.[52] In Illustration 8, Walker used Figure 22 to explain that a surplus of electricity inside the jar forces electricity away from the outer surface to produce a deficit.

When the inner and the outer surfaces are connected by a metallic conductor, sparks fly as the internal excess electricity rushes between the adjacent knobs *a* and *e* to restore equilibrium.

Many electrical philosophers eagerly adopted Franklin's suggestions, which provided a common conceptual basis for decades. Particularly in France and Italy, theoreticians developed his ideas further, incorporating them within revamped models as part of their own bids for intellectual power. But, like his opponents, Franklin was forced to shore up his system with specially devised hypotheses that worked marvellously in some instances but completely failed in others. He never did manage to explain why negatively charged bodies (in his system, those with a deficit of electrical matter) repel each other. Before the end of the century, several researchers had independently rejected atmospheres and aethers, and begun to think about electricity's actions in terms of forces.

# 3    Theological Aethers

The belief that some sort of aether pervades the universe had been prevalent since the time of Aristotle. Over the centuries, writers postulated aethers with various characteristics that would simultaneously explain different physical phenomena and also comply with existing philosophical and religious beliefs. During the second half of the eighteenth century, electrical aethers proliferated. There is, however, no simple way of classifying them. Some, like Franklin's, originated from Newton's suggestions, some were based on Descartes' hypotheses, while several borrowed features from both sources. Many aethers consisted of small particles with special properties, but others – Nollet's, for instance – contained a fiery material. And many were so nebulously defined that it is hard to determine what their authors meant.

Diverse though they were, electrical aethers did hold several characteristics in common. They were invisible and weightless; they failed to explain all of the experimental observations; and they were only vaguely described. Instead of examining their physical properties, it is more rewarding to explore their theological significance. Modern scientists often claim that their religious beliefs and their scientific work are completely separate, but experimental philosophy originated in attempts to reconcile biblical accounts with evidence drawn from the natural world.

Especially in Britain, natural theologians preached that the fundamental purpose of natural philosophy was to celebrate God. Uncovering Nature's laws revealed God the Designer, who had created a wonderful universe for people to live in and care for, a dominion that He ruled wisely and omnisciently. One major function of electrical aethers was to account for God's continuing involvement in the world. As an Essex doctor put it: '*Æther* is the *Rudder* of the Universe, or as the *Rod*, or whatever you will liken it to, in the *Hand* of the *Almighty*, by which he *naturally* rules and governs all *material* created *Beings* ... Now how beautiful is this *Contrivance* in God.'[53]

Most writers in the eighteenth century held a dualist position: they believed that there are two distinct substances, mind and matter, so that God, angels and human souls are fundamentally different from physical objects. Many people objected that Newton's notion of particles attracting one another

117

through empty space was heretical, because it attributed spiritual power to inert matter. On the other hand, dualism immediately raised several problems. How can spiritual and material entities interact with one another? How can the brain control the body? And how can God influence what happens on earth? Aethers provided a convenient intermediary, since they were – like God – invisible, extended throughout apparently empty space, and could act on inert matter.

Many of the aethers intended to fulfil this intermediary role were derived from a tentative suggestion made by Newton:

*And so if any one should suppose that Æther (like our Air) may contain Particles which endeavour to recede from one another (for I do not know what this Æther is) and that its Particles are exceedingly smaller than those of Air, or even than those of Light: The exceeding smallness of its Particles may contribute to the greatness of the force by which those Particles may recede from one another, and thereby make that Medium exceedingly more rare and elastick than Air; and by consequence ... more able to press upon gross Bodies, by endeavouring to expand it self.*[54]

Newton had disguised this radical proposal as one of several '*Quaeries*' at the end of his book *Opticks*. These queries, which ranged over electricity, chemistry and many other topics, provided an experimental agenda at the Royal Society for much of the eighteenth century.

At first, natural philosophers mainly focused on Newton's suggestions about the short-range forces between tiny particles. But from around 1740, Newtonian aethers became increasingly popular. Different aether models were devised to explain a wide variety of phenomena, including gravitational attraction, the body's nervous functions, and optical effects. Franklin's electrical fluid was typical of these neo-Newtonian aethers. In his system, there are two kinds of particles. Ordinary ones are attracted towards each other to form water, gold, amber and so on, while special minute particles make up an electrical aether. Although these subtle aetherial corpuscles repel one another, they attract 'gross Particles' (meaning large as well as base) to different extents, so that electrics such as glass retain electrical matter far more strongly than non-electrics, notably metals.

Aethers like Franklin's thus performed two functions: they gave a moderately satisfactory explanation of many – but not all – electrical experiments; and – provided they were not examined too closely – they offered a reasonable account of how Christian dualism and electrical theory could be reconciled. Yet, although this compromise was broadly acceptable, there were many dissenters.

Using theological arguments, natural philosophers suggested other forms of aethers that behaved very differently from Franklin's. Descartes and the mechanical philosophers had insisted that, since matter is inert, motion can happen only through direct contact. During the eighteenth century, some deeply religious scholars turned to the Bible for evidence to support their view that the world must be packed full, with no empty spaces. Although they were often dismissed as cranks, they attracted small cliques of influential followers. The most important were the Hutchinsonians, a group whose members clandestinely propagated the ideas of John Hutchinson, a fervent anti-Newtonian. Because Newton has acquired such heroic proportions in science's history, it is easy to forget that during the eighteenth century, his protagonists were still struggling to assert their supremacy over opponents like the Hutchinsonians whose voices have effectively been silenced.

In Hutchinsonian cosmologies, the universe operates like a huge machine propelled by a constantly circulating aether originating from the sun. God set this perpetual machine in motion at the Creation, yet can interfere sporadically by performing miracles. Electrical effects are due to the mechanical pressure of a subtle aether, which drives electrified bodies towards one another or else pushes them apart.

Ironically, although such aethers were religiously inspired, they sounded dangerously like the mechanical models of materialists who, because they attributed attractive and repulsive powers to matter itself, were often denounced as atheists. Another approach was to make the aether itself inherently active. This was the route taken by Bishop George Berkeley. He is now remembered mainly as a philosopher, but during his lifetime he was renowned for his optical ideas, his medicinal tar water, and his theological objections to Newton's natural philosophy and mathematics. Berkeley's aether was a divine and animate entity, what he described as 'the vegetative soul or vital spirit of the world'.[55]

Berkeley had derived his aether by drawing on classical traditions as well as contemporary chemical discussions about the nature of fire. During the second half of the eighteenth century, other writers in their turn modified Berkeley's model, either directly or through inheriting variations devised by other writers. Adam Walker, for instance, envisaged a solar aetherial fire constantly circulating the universe and striving to achieve equilibrium. Electricity, light and fire were, he taught, all modifications of the same aetherial principle.

Other devout proponents of spiritual aethers were strongly influenced by their Trinitarian faith. For them, there were close parallels between the three aethers that they used to explain heat, light and electricity, and the Christian belief in God's triple manifestation as the Father, Son and Holy Spirit. Just 'as the mechanic philosophers make the Aether the cause of attraction, muscular motion and other extraordinary phænomena of matter: So is the HOLY GHOST the cause of all spiritual conduct, which is consonant to the divine Law.'[56]

Not all natural philosophers chose aetherial explanations for electricity, and towards the end of the century, arguments against invoking these unobservable fluids became increasingly dominant. Nevertheless, as theological aethers were repeatedly tailored to fit other theoretical positions, their influence continued well into the nineteenth century. In particular, they profoundly affected the thought of Michael Faraday, who unified electricity and magnetism into a single field model. Although nineteenth-century scientists were less explicit about the theological foundations of the new aethers that they invented, many of them were as concerned as their Enlightenment predecessors to make them compatible with Christian beliefs.

# 4 Measurement and Mathematics

Like Newton, his intellectual hero, Joseph Priestley disguised his more speculative suggestions as questions. Right at the end of his *History and Present State of Electricity*, Priestley examined the behaviour of pith balls in an electrified metal jar and ingenuously enquired: 'May we not infer . . . that the attraction of electricity is subject to the same laws with that of gravitation, and is therefore according to the squares of the distance . . .?'[57] Priestley's deceptively casual remark could have been made by many Enlightenment philosophers. Despite their claims of neutrality, they often embarked on their investigations convinced in advance that Newton's formula for gravitational attraction governs the activity of physical, chemical and biological systems.

Electricity was no exception. By exploring manuscript archives, historians have discovered that several British researchers successfully performed electrical experiments that were designed not to test but to confirm their Newtonian preconceptions. However, it was a French military engineer, Charles

Coulomb, who was the first to make his results public knowledge. It is Coulomb, therefore, who has become acclaimed as one of electrical science's most famous pioneers, and whose name is now used as a unit of electric charge.

Born in 1736, Coulomb belonged to a new generation of natural philosophers whose approach predominated in the last third of the century. Rather than accumulating qualitative descriptions of electrical effects, they made accurate measurements and tried to organise their discoveries into systematic patterns. Instead of relying on the evidence of their own bodies, they invented delicate and precise instruments. Rejecting unobservable aethers and atmospheres, they preferred to think in terms of forces that could be quantified and analysed mathematically. These shifts in electrical research were typical of changes taking place at different rates in many experimental fields during the later Enlightenment period.

Coulomb was a product of the French educational system, which was strongly oriented towards mathematics and engineering. Unlike in England, where experimenters had to find their own funds, state intervention stimulated theoretical research and fostered technical improvements. Coulomb embarked on his first major project in response to a competition set by the Paris Academy. Typically, the topic that was set carried important practical applications. Contestants were asked to invent a better compass for investigating the earth's magnetic patterns, a project intended to improve navigational safety.

Benefiting from years of engineering experience acquired abroad during warfare against England, Coulomb designed an almost frictionless compass by suspending the magnetised needle from a fine silk thread. It was soon installed to make accurate measurements at the Paris Observatory. Unfortunately, this instrument seemed to be *too* sensitive, quivering at the slightest change in its surroundings. After diagnosing electrical interference, Coulomb investigated the properties of piano wire, modified his compass, and then used the new skills he had gained to examine electricity.

Coulomb established his international reputation by successfully using a new instrument, a torsion balance, to show that electrical effects can be described by a Newtonian inverse square law. His torsion balance, shown in Illustration 12, relies on a simple yet ingenious principle: it measures how strongly two light elder balls are electrostatically pushed apart by observing the angle through which a wire must be twisted to hold them at a certain distance.

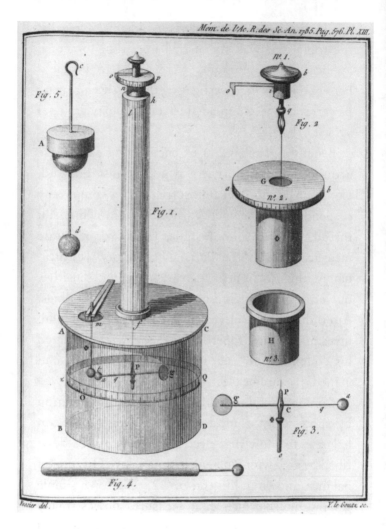

*Illustration 12*: Charles Coulomb's torsion balance.
*Histoire de l'Académie Royale des Sciences*, 1785, Plate
XIII, facing p. 576. (Cambridge University Library)

From the outside – his Figure 1 – Coulomb's apparatus (which was about 90 cm high) appears as two glass cylinders on top of one another. Inside, a fine silver wire hangs down through both cylinders and carries the device shown in Figure 3, a light horizontal rod with a ball *a* at one end, and a counterbalancing paper disc *g* at the other. From the top of the lower cylinder, the second ball drops down through a hole so that it can be electrified by the probe in Figure 4. After electrification, the distance between the balls is recorded by the scale **OQ** pasted to the outside of the lower cylinder, while the torsion on the thread – the force due to its being twisted – is measured at the top of the wire, as illustrated in Figure 2. By knowing the torsion characteristics of the wire, Coulomb could use a simple formula to work out the force between the balls.

Straightforward in principle, the experiment was awkward to perform: his piano wire kept twisting round itself, and some of his measurements deviated far from their expected values. Nevertheless, Coulomb confidently declared that the balls obeyed a Newtonian inverse square law. Verifying the relationship for attraction proved even more tricky, because the balls kept sticking to one other, so he devised a different apparatus. Although it was intrinsically less accurate, Coulomb managed to explain away observational discrepancies, and he triumphantly confirmed that the attractive force between oppositely charged balls is inversely proportional to the square of the distance between them. Although Coulomb's experiment was difficult to replicate, his conclusions were immediately welcomed in France, and within twenty years had become acknowledged as electrical truth all over Europe.

One of the few experts cited by Coulomb was the German theoretician Franz Æpinus, who in 1759 had published a sophisticated mathematical treatment of electricity and magnetism based on Franklin's qualitative one-fluid model. Writing in Latin, and isolated in his academic outpost in St Petersburg, Æpinus had found it difficult to find sympathetic readers for his new algebraic analyses. With Coulomb's influential endorsement, Æpinus's work became far better known.

Because he was primarily interested in magnetism, Æpinus dealt only with those electrical phenomena that were analagous to magnetism. But although not a systematic treatment of electricity, it was extremely influential. At the outset, Æpinus made his debt to Franklin clear, yet he soon deviated. Unlike Franklin, he insisted that impermeability to electrical fluid is a property shared by all insulators, not one specific to glass. Still more importantly, Æpinus probed more deeply the concept of a fluid. By dispensing with flow, and focusing on the fluid's influence through its presence, Æpinus opened up the possibility of developing a theory based on action at a distance.

As the first major mathematical analysis of electricity based on forces, Æpinus's *Essay* stimulated new theoretical and experimental methods. By the end of the century, following Coulomb's endorsement, a distinctive French school of electricity had developed. As in other areas of research, it was characterised by a rigorous mathematical approach, precise experimental measurement, and a determination to explain effects in terms of simple Newtonian laws of force.

The situation was very different in England. London's instrument makers were the finest in the world, yet they were competing to market their apparatus rather than to provide objective measurements or test new theories. Furthermore, scarcely any men of science possessed the mathematical skills needed to appreciate Æpinus's innovations. When Benjamin Wilson received his copy, he politely initiated Æpinus into local customs: 'The introducing of algebra in experimental philosophy is very much laid aside with us, as few people understand it; and those who do, rather chose to avoid that close kind of attention: tho' I make no doubt but I dare say you had a very good reason for making use of that method.'[58]

In Britain, even those men who were capable of tackling Æpinus developed electrical theory along geometrical rather than algebraical lines, insisting that this was the only valid way to represent physical phenomena. George Adams (originator of Illustration 11, page 89) explained his antagonism to mathematical theorising: 'Philosophy owes much to the assistance it has received from mathematicians; but this only happens when they apply themselves to the study of phenomena; when neglecting those, calculations are made to serve an hypothesis; the more elegant and beautiful they are, the more detrimental they become to science. It is thus that *Æpinus*, by a mathematical theory of electricity, has closed the door on all our researches into the nature and operation of this fluid.'[59]

Because algebra was perceived as a French innovation, debates about using mathematics in electrical research were imbued with political interests. In 1801, a special supplement of the *Encyclopædia Britannica* was dedicated to preventing 'the seeds of Anarchy and Atheism' reaching England from post-Revolutionary France; beware, warned one article, of the French algebraists who manipulate meaningless symbols like 'a banker's clerk . . . this total absence of ideas, exposes even the most eminent analyst to frequent risks of paralogism and physical absurdity'.[60]

The British hostility towards algebra and calculus was due neither to stupidity nor laziness, but stemmed from a variety of historical, educational and philosophical grounds. This aversion persisted for many decades, and even Faraday – Britain's leading electromagnetic scientist – spurned the continental mathematical techniques that now seem inseparable from scientific thought. Just as had happened during the Enlightenment, the sort of electrical science carried out in the nineteenth century varied from country to country.

# Part V
# The Flow of Life:
# Current Electricity

*This endless circulation of the electric fluid (this perpetual motion) may appear paradoxical and even inexplicable, but it is no less true and real; and you feel it, as I may say, with your hands.*
  Alessandro Volta, *Philosophical Magazine*, 1800

Few people now recognise the name of Johann Sulzer, a German mathematics professor who unwittingly passed over his chance for posthumous glory as the discoverer of current electricity. Like many of his Enlightenment contemporaries, Sulzer used his own bodily senses as investigative instruments. One day in around 1750, he sandwiched his tongue between two pieces of metal, one of zinc and one of copper, and noticed a pungent taste that reminded him of green vitriol (iron sulphate). Unfortunately for his posthumous fame, he chose to ignore this intriguing finding, and almost half a century went by before similar taste experiences helped to decide between competing electrical theories.

It 'can hardly be too often repeated', preached Priestley, 'that more is owing to what we call *chance*, that is, philosophically speaking, to the observation of *events arising from unknown causes*, than to any proper *design*, or pre-conceived *theory* in this business'.[61] The history of current electricity is littered with accidental discoveries, discarded explanations, and arguments won by persuasion rather than reason. Because Alessandro Volta's pile became the prototype for batteries capable of producing a continuous flow of electricity, it is often heralded as the first triumph of nineteenth-century physics. We celebrate Volta rather than Sulzer: he is an Italian national hero, and volts are international units of measurement. Volta himself was, of course, unaware that his invention would lead eventually to the modern electrical industry. He had developed it in order to win a bitter controversy that had raged throughout the 1790s about Luigi Galvani's claims to have discovered a new type of electricity.

This power of serendipity is also central to the most famous anecdote about Volta's rival, Galvani, now remembered in expressions like 'galvanised steel' or 'galvanised with shock'. Apparently, when

Galvani accidentally touched the nerve in a frog's leg he was dissecting, he noticed that it twitched in unison with a sparking electrical machine. This coincidence was perhaps more carefully contrived than Galvani would have us believe, but he dedicated much of the rest of his life to exploring this supposedly chance event.

Confronted with a plethora of unconnected observations, even sober natural philosophers indulged in fanciful speculation. Perhaps, mused Priestley, it was significant that the feathers of thirsty parakeets become electrified; or, he continued, could the brain be a processing laboratory for invisible fluids, busily converting phlogiston into electricity? After all, since living beings responded so strongly to electricity, it did not seem unreasonable to suggest that they might be able to generate it. Electrical research was carried out not in the large organised physics laboratories that developed during the nineteenth century, but in animal dissection rooms, hospital clinics – and even on fishing trips.

The nature of animal electricity lay at the heart of one of science's most famous debates, one which has often been picturesquely portrayed as a personal battle between two hot-headed Italians, Volta and Galvani. But far more was at stake than their individual reputations: this dispute resonated throughout

Europe because it was also about the nature of life itself. Are living organisms powered by electricity? And could animal electricity be identical with the electricity produced by machines?

Electricity is now regarded with respect as a deadly power, one that has been safely harnessed for domestic and industrial use. But in the eighteenth century, electricity was seen as a life-giving force. As Enlightenment philosophers experimented with the static electricity produced by electrical machines and Leyden jars, even the rashest optimists had no inkling that in the future, current electricity would be running trains, illuminating buildings and energising computers.

# 1    Henry Cavendish and the Torpedo

Although Henry Cavendish (1731–1810) is not famous, he was one of Britain's richest and cleverest men at the end of the eighteenth century. The frugal inheritor of a massive family fortune, this scion of the Dukes of Devonshire renounced two ancient family traditions: engaging in political activities, and residing in opulent country mansions. Instead, Cavendish chose to live modestly in central London, where he devoted himself to experimental research over a wide range of fields.

Like many unusual people, Cavendish has been interpreted in various ways. Victorian biographers seized on his reclusiveness, his penchant for shabby clothes and his inherited physical awkwardness to romanticise about a tortured genius devoid of ordinary human passions. Many historians still portray him as a great British eccentric, recycling the pithy comment made by one of Volta's friends that Cavendish was 'a man so unsociable and cynical that he could stand honourably in the same tub with Diogenes'. Recently, more sympathetic psychobiographers have reappraised his colleagues' opinions to diagnose a brilliant yet excessively shy man, probably afflicted with chronic depression, who courageously struggled with his personality problems to participate in the affairs of the Royal Society.[62]

Cavendish dedicated his life to scientific investigation, yet published very little. Researchers are still sifting through the voluminous boxes of hand-written manuscripts that have survived, and whose content provides evidence of his extraordinary abilities. It has become clear that one of his mourners was not exaggerating when he wrote: 'You will have heard that we have lost Mr. Cavendish, – a man of a wonderful mind, more nearly approaching that of Newton than perhaps any individual in this country since his time.'[63] Newton was the inspiration for what is now Cavendish's most famous experiment, his adaptation of Coulomb's electrostatic apparatus (Illustration 12, on page 126) to measure gravitational force, and hence to weigh the world – a dramatic label that conveys little of the skill demanded by this feat of precision measurement.

Cavendish was – among many other things – an expert on electricity. One of the few Englishmen capable of understanding and formulating mathematical electrical theories, he was also a superb experimenter. He was recruited for the committee that investigated how effectively lightning rods could protect the nation's gunpowder stores (Illustration 9, on page 80), and was renowned throughout Europe for his investigations of electric fishes, whose behaviour became a subject of keen interest.

In the 1770s, reports came in from South America that a giant eel could deliver electric shocks powerful enough to slay men and horses. Could this eel perhaps be related to the torpedo, a flat ray fish named after its mysterious ability to numb its prey into a torpor? Many people still regarded the torpedo as a magical creature, probably related to mythical submarine monsters that could suck ships to a halt. Rational philosophers took great pride in discrediting such beliefs, which in this case stemmed from a rich literary tradition stretching back to Pliny, the first-century Roman naturalist.

Rather than studying South American eels, English researchers set out to investigate the torpedo, which was conveniently available for study in European waters, and show that it too operated through electricity rather than by exerting an occult power. One of the most energetic enthusiasts was John Walsh, a Member of Parliament who had formerly been Robert Clive's secretary in India. Benefiting from the expertise of seasoned fishermen, he examined several different kinds of torpedo under various conditions, carefully recording their size, as well as the strength and frequency of their shocking capacities.

Convinced that their behaviour could be explained electrically, Walsh enlisted allies to support his cause and persuade sceptics that the torpedo acted like a set of Leyden jars. He engaged one of London's leading surgeons to dissect and draw the fish, which he persuasively rechristened 'the Electric Ray'. For publication in the *Philosophical Transactions*, Walsh wrote several letters to Franklin. His sycophantic capital letters almost deified the doyen of electricity: 'I rejoice in addressing these communications to You. He, who predicted and shewed that electricity wings the formidable bolt of the atmosphere, will hear with attention, that in the deep it speeds an humbler bolt, silent and invisible . . . He, who by Reason became an electrician, will hear with reverence of an instinctive electrician, gifted in his birth with a wonderful apparatus, and with the skill to use it.'[64]

Gently mocking traditional beliefs that a torpedo could cure headaches and gout, Walsh prescribed instead a dose of electricity, advising invalids 'that *the Leyden phial contains all his magic power*'.[65] 'Torpedinal electricity', he argued, was generated by a pair of special electrical organs, each divided into hundreds of columns separated into tiny, liquid-filled spaces. Packed with nerves to create and control the ray's electric fluid, these functioned like large batteries of Leyden jars connected together. But not everyone agreed with Walsh's bold claims. As he was forced to admit, a torpedo could not produce a spark or make pith balls spring apart, two basic properties of electricity. And how did it manage to deliver a shock under water, when – as was believed – it was easier for electricity to travel through water than through a human body? To resolve these problems, Cavendish decided to build an artificial torpedo that would replicate the living fish's performance and prove that it did, indeed, deserve to be called an 'Electric Ray'.

After some trial runs, Cavendish concluded that pieces of thick leather, soaked in salt water and cut out into a fish-like shape, provided the best material for the torpedo's body. In Illustration 13, Figure 3 shows one side of Cavendish's symmetrical apparatus. Thin plates of pewter Rr, designed to simulate the torpedo's electric organs, were attached to a battery of forty-nine Leyden jars by insulated wires Ww passing through a long glass tube MNmn. Apart from the long handle, the whole assemblage was covered with sheepskin, and could be immersed in

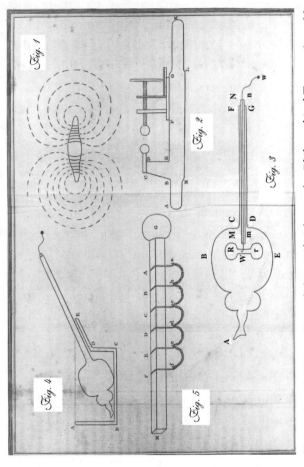

*Illustration 13:* Henry Cavendish's artificial torpedo. *Philosophical Transactions 66,* 1776, facing p. 222. (Cambridge University Library)

salt water (Figure 4). Cavendish explained that since electric fluid (the dotted lines in Figure 1) constantly circulated through the torpedo (the central solid outline), anyone touching the fish would receive a shock, which would be felt especially strongly near the electrical organs.

Cavendish's first step was to remove one of the key objections to his claims by studying how the emission of sparks was affected by the construction of the battery. Using a hand-held version of Timothy Lane's electrometer (Illustration 11, on page 89, and Illustration 13, Figure 2), he found that the more jars there were in a battery, the shorter the distance a spark could travel. Since, he argued, the torpedo's special organs contained hundreds of tiny cells, each analogous to a miniature Leyden jar, it was quite plausible that these could collectively deliver a substantial shock, yet be unable to eject the electric fluid with enough force to produce a visible spark. Cavendish also quashed the pith ball challenge, maintaining that the torpedo completes its discharge so quickly that balls could not have time to separate.

Presumably conquering his aversion to social encounters, Cavendish invited many distinguished visitors to test and admire his artificial torpedo, carefully noting their individual reactions. Priestley, for instance, bore a shock up to his elbow, and even sceptics went away converted by their painful experiences. Cautiously concluding only that his artificial torpedo demonstrated how its living counterparts *could* be electric, Cavendish carried out some nifty arithmetic to prove that they could pack fourteen times as much electricity into their bodies as his battery of Leyden jars.

In the debates about animal electricity that raged across Europe for the next quarter of a century, Cavendish's experiments were repeatedly cited. Sceptics remained unconvinced that torpedoes were electric, since it did not seem possible that an instrument built from glass and metal could function in the same way as a living organism. To many doubters, it verged on sacrilege to suggest that a man-made machine, which generated electricity artificially, could perform in the same way as a natural being created by God. It was this equivalence of animal and artificial electricity that lay at the heart of the confrontation between Galvani and Volta.

# 2   Luigi Galvani and his Frogs

'The Birth of Wonders!', exclaimed the anonymous *Sceptic* in 1800, satirically explaining that the spirit of confusion had 'fled into Italy, and entered into the body of a Frog, which professor Galvani was dressing for his Wife's supper'.[66] Galvani (1737–98) had died two years earlier, just before his final defeat by Volta in the battle about their electrical theories. Yet all over Europe, people were still discussing his frog experiments, which are famous today because he discovered how an electric current can be produced by bringing together two different metals.

Galvani's electrical investigations make grisly reading, yet he was no more cruel than his predecessors. Eighteenth-century researchers apparently had few qualms about subjecting live creatures to electrical ordeals. An Italian experimenter 'arranged for one of the extensile muscles of a live rooster's leg to be separated from the thigh, leaving it attached to the leg by tendons and nerves alone. I fastened a brass wire to each tendon . . .', and he then passed electric shocks through it. However, *salon* life was not far away: he coolly reported that the pieces of muscle swelled and moved apart, 'resembling a lady's fan being opened briskly'.[67]

Italian physicians were also Europe's leading exponents of electrical medicine. In 1759 in Bologna, Galvani's own city, a group of spectators admired the successful treatment of a man who had been paralysed for two years. 'It was a fine sight to see the mastoid rotate the head ... the biceps bend the elbow; in short, to see the force and vitality of all the motions occurring in every paralysed muscle exposed to the stimulus.'[68] Roosters and people: for his work with frogs, Galvani had many local precedents to draw on.

Galvani initially specialised in anatomy rather than electricity. Trained in medicine, his early papers were on esoteric subjects such as birds' kidneys. Galvani eventually became a professor of obstetrics, a career rise perhaps helped by marrying the daughter of Bologna's leading anatomy professor, who became his indispensable research assistant. They were so devoted that he even occasionally referred to her as 'my wife' in his private experimental reports, instead of concealing her presence behind the usual veil of anonymity. Galvani was devastated when she died after almost thirty years of marriage, and his life went into a sad decline.

Because he refused to swear an oath of allegiance to Napoleon's Cisalpine Republic, he lost his academic posts and died an impoverished, depressed man.

Galvani was already in his early forties when he first started working on electricity. In retrospect, this seems a sensible switch of interest. Italian universities were starting to pay their lecturers higher salaries, so that Italy was gaining the reputation of replacing Britain as the best place for scientific studies. Moreover, investigators throughout Europe were focusing on the relationships between life, movement and feeling. How, they asked, does the brain interact with its body's environment, sending signals to the muscles instructing them to move, and receiving information back from the sense organs? Philosophers had been debating this problem since antiquity, but physiologists had turned to it with renewed vigour in the eighteenth century. Now it seemed that the answer might lie in electrical experiments. Galvani himself certainly hoped for electrical enlightenment, predicting that his experiments would 'shed some light on the darkness still shrouding the phenomena of nerves'.[69]

The traditional explanation for motion and sensation had been that a nervous fluid, often called 'animal spirits', transmits messages between the brain and the body by passing swiftly through extremely fine canals. But from around the 1740s, natural philosophers adapted Newton's conjectures about an aether to suggest that sensations are carried by the vibrations rather than the flow of some sort of special nervous medium. The most recent theory identified electricity with animal spirits. This solution was much contested, but immediately overcame some of the obvious objections to the two older models. Electricity moved extremely rapidly, and could travel through animals with dramatic results. Moreover, argued the proponents of animal electricity, this theory was strengthened by torpedo studies, and by laboratory experiments showing that electricity could cause the muscles of dead animals to contract.

So Galvani's innovation was not to invent the concept of animal electricity, but to provide new experimental evidence of its existence. By 1780, he already had considerable experience of working with frogs, since for several years he had been examining their muscles and the effects of opiates on their nerves. According to his own account, his research took a dramatically new direction when he performed some experiments on a dead frog that happened to be near an electrical machine (Illustration 14). The dissected frog was lying on a Franklin square, a plate of glass with metal foil on

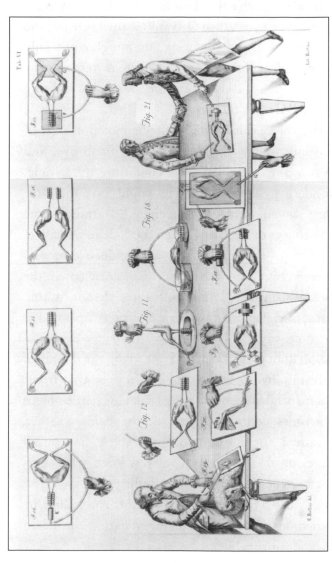

*Illustration 14:* Galvani's frogs. Luigi Galvani, *Opere Edite et Inedite*, Bologna, 1839, Plate VI. (Cambridge University Library)

each side, a device that Franklin had designed to act like a Leyden jar. When Galvani's scalpel touched the inner nerve in the frog's leg, its muscles not only contracted violently – as if it had tetanus, he commented – but also kept time with sparks being discharged from the machine.

To explore this bewildering phenomenon further, Galvani carefully repeated the experiment over and over again, changing one condition at a time. Wondering whether lightning would have the same effect as an electrical machine, he attached the earthed legs of frogs and sheep (both dead and alive) to long iron wires during a thunderstorm, and repeated his earlier tests. As he had anticipated, the muscles contracted just as before, demonstrating that artificial and atmospheric electricity operated in the same way.

Boredom played a crucial role in the third stage of Galvani's investigations. He had noticed that frogs' legs suspended by brass hooks from his garden fence sometimes contracted even during calm weather. Impatient for more frequent results, he started squeezing the hooks and was rewarded with sporadic twitches. Eventually concluding that the effect arose when a brass hook touched the iron railing, he carried out more systematic investigations inside on his laboratory table.

Placing frogs on plates made of different materials – metals, glass, resin – Galvani determined the conditions under which the convulsions could be reliably obtained. Illustration 14 shows how he varied his basic experiment to explore the startling new phenomenon he had discovered. He found that when he touched one end of a curved metal conductor to the brass hook in a frog's spinal cord, and the other end to its leg, the muscles would contract. In the centre, Figure 11 depicts a macabre dancing pendulum. The disembodied hand (perhaps his wife's) is holding one leg of a frog with its brass hook touching a silver plate. As soon as the other leg touches the plate, its muscles contract so that it rises up into the air. But as soon as the contact is broken, the muscles relax, the leg falls, and the whole cycle starts again.

From a modern vantage point, Galvani's results were exciting because he had demonstrated that an electric current is produced when two different metals touch each other in the moist, conducting environment of a frog. But Galvani interpreted his findings differently. For him, they confirmed the existence of animal electricity. Drawing mainly on evidence from frogs and torpedoes, he claimed that all animals possess a special electric fluid that is generated in the brain and passes through the nerves into the muscles, which function like miniature Leyden jars.

A meticulous researcher, Galvani kept systematic records of his frog discoveries. As well as maintaining a daily journal, he wrote several complete manuscripts that were never published, and he had fully developed his ideas several years before his Latin *Commentary on the Effects of Electricity on Muscular Motion* finally appeared in 1791. Could he have been suffering some sort of psychological block? Certainly the tone of his writing is – in striking contrast to Volta's – cautious and conjectural. He ordered only twelve copies for the first edition, and presented one of them to Volta. Perhaps Galvani delayed publication because he feared the consequence that did, in fact, transpire. After Volta's initial enthusiasm had soured to scepticism, he energetically deployed diverse tactics to promote himself and publicly disparage the diffident, conscientious anatomist from Bologna.

# 3   Alessandro Volta and his Pile

Volta (1745–1827) was a flamboyant man who thrived on controversy. He was also one of the first men to embark on a career in physics, an early example of those paid professionals who later in the nineteenth century would become known as scientists. To consolidate his status, Volta made sure that his work was discussed internationally, even though he was operating in Pavia, far from the major centres of Paris and London. He became involved in a variety of projects, but by far his most famous invention was his pile, the forerunner of electric batteries that generate current. Launched in 1800, two years after Galvani's death, Volta's pile is often said to be the last Enlightenment instrument, one that marked not only a new century, but also the beginning of modern physics.

Although Volta came from a distinguished clerical background, and remained a devout Catholic throughout his life, he enjoyed a reputation for womanising, and the polemical style of his articles makes it clear why his family wanted him to study law. But Volta refused: his genius, he insisted, directed him towards electricity. Even as a teenager, he corresponded with Europe's leading electrical authorities, and he soon started devising instruments and experiments of his own. By the time he was thirty-two, he had persuaded the Austrian government (which then controlled his home duchy of Milan) to give him a travel grant to visit experts abroad. The following year he was appointed the professor of experimental physics at the University of Pavia, a post he held for forty years.

At Pavia, he used state money to improve the University's teaching facilities, and also to finance his own trips for establishing valuable personal contacts. At the same time, he strengthened his reputation abroad by cultivating friendships with prominent investigators such as Priestley. Volta's scientific battle with Galvani coincided with military warfare, as Napoleon extended his empire over Italy. Unlike Galvani, after displaying some initial resistance, Volta diplomatically acknowledged the new regime. This political accommodation later paid off, as Napoleon became Volta's most influential admirer, rewarding him with a pension as well as administrative power.

When Volta received his copy of Galvani's *Commentary*, he was already renowned for his innovations in electricity, meteorology and pneumatics (the study of gases). Within only a few weeks of learning about the Bologna experiments, Volta had begun to criticise them. Galvanic fluid, insisted Volta, was no different from the electricity produced by electrical machines; it originated not in a frog's nerves – as Galvani contended – but in the contact between the metals applied to its leg. Europe's electricians divided into different camps, as they discussed three basic questions. Was Galvani's fluid electrical or not? Did it originate outside or inside the experimental frogs and torpedoes? And was it the same as nervous fluid?

Whereas Galvani had focused on the frog to explore what makes an animal different from inert matter, Volta viewed the frog as a sensitive instrument, one that could reveal the similarities between living creatures and the inanimate world. To resolve the contest, Volta chose a stratagem that he had already used successfully: he decided to develop a new instrument. This is often a powerful move because an impressive experiment deflects attention away from fundamental theoretical issues, and forces opponents into a defensive position. In 1800, after almost a decade of research interrupted by University commitments, his growing family, and temporary exile to avoid harassment during the

French occupation, Volta announced his pile to the world.

Of course, the fact that an instrument works does not prove the truth of the theory on which it is based. Nevertheless, Volta dazzled Europe's electricians with his new pile. He once asserted that 'the language of experiment is more authoritative than any reasoning: facts can destroy our ratiocination – not vice versa'.[70] This maxim sounds like one of the basic rules for conducting science, yet Volta was a skilled rhetorician who won over his audiences through force of argument rather than incontrovertible physical evidence. Victory in battle depends on the strength of one's allies, and Volta was far more effective than his opponent at soliciting support. By targeting the right people, Volta persuaded the scientific world that his pile refuted Galvani's demonstration of animal electricity.

The basic construction of Volta's pile is shown in Figure 2 of Illustration 15. It consists of a column of discs made from two different types of metal (often zinc and silver) each roughly an inch in diameter, piled up on top of one another and separated by pieces of cardboard soaked in salty water. The vertical glass or wooden rods (*m*) prevent the pile from toppling over. Volta explained that if the experimenter dips his fingers into the basin of water

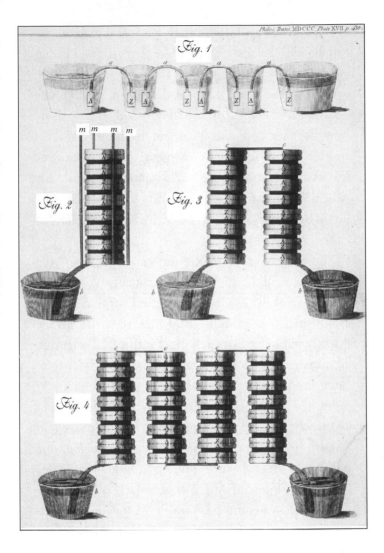

*Illustration 15*: Alessandro Volta's pile. *The Philosophical Magazine* 7, 1800, p. 336. (Cambridge University Library)

at the bottom, and then touches a metal plate at the top, he will feel a shock just like that from a Leyden jar or a torpedo. To obtain stronger effects, Volta wanted to use more pairs of discs, but his columns became unmanageably tall; Figures 3 and 4 show how he overcame this problem by joining several shorter ones together with metal plates. Figure 1 shows a variation called the 'chain of cups', for which Volta connected up to sixty glass tumblers of liquid by metallic arcs plated with a different metal at either end.

To achieve the maximum impact from the announcement of his new instrument, Volta took advantage of the English connections he had already established. In March 1800, he sent his article directly to one of Europe's most influential scientific administrators, Sir Joseph Banks, President of London's Royal Society. Almost immediately, it was published in the original French in the *Philosophical Transactions*, and by September had appeared in English in the smaller *Philosophical Magazine*, one of the new journals springing up in England to cater for a wider, if less prestigious, scholarly community.

From the opening paragraph, Volta's paper conveys a sense of excitement and is designed to impress his distant audience. Banks and the Royal Society will, Volta informs them, learn about 'striking results' from an 'apparatus . . . which will, no doubt, astonish you'. He then proceeds to bombard his readers with startling discoveries, deceptively relating his own personal experiences to give the impression that he is candidly revealing the progress of his investigations. So we learn about the degrees of shock he felt with different experimental configurations, and how he used his own body as an instrument to explore the effects of electricity on his sense organs – taste, sight, hearing and smell, as well as touch and pain.

Whereas Sulzer had discarded his accidental taste sensations, Volta explored them systematically. Like Galvani, he converted an apparently chance observation into a powerful argument, but used his 'to combat the pretended animal electricity of Galvani, and to declare it an external electricity moved by the mutual contact of metals of different kinds'.[71] By placing two different metal plates on either side of his tongue, Volta transformed himself into the human equivalent of an experimental frog – but one that could speak as well as twitch. The sensations in his mouth, Volta reported, depended not only on the size of the pile, but also on which way round the plates were placed. Thus, he maintained, the electricity came not from his tongue but from the metals.

Volta's forceful article leaves no room for

questions. Apparently casual asides conceal new theoretical explanations that undermine the concept of animal electricity, even though the phrase itself rarely appears. At the end of his breathless account, Volta confidently boasts that his instrument confirms not only that electricity is produced by metallic contact, but also that his artificial pile is 'at bottom the same as the natural organ of the torpedo'.[72]

Packed with observations and conjectures dressed up as assertions, Volta's letter is also remarkable for what it leaves out. Although appearing honest and detailed, it includes no precise instructions about the best type of metal and liquid to use. Even the diagrams would only be of limited use for someone trying to replicate his apparatus. Most significantly, Volta fails to mention the gas bubbles and other chemical activity that he must have noticed, and also glosses over the short duration of his pile's effectiveness. Discussing these factors would have focused attention away from the metals and onto the liquid, which is crucial for explaining the pile's operation but not for justifying Volta's claims.

As well as vital experimental details, Volta omitted alternative theoretical explanations. Introducing new terms such as 'electro-motive' and 'perpetual motion', he maintained that his observations provided 'proof' for his ideas. However, he was not *testing* his theory, but instead was setting out to *confirm* it by suppressing unwanted evidence. There is also a missing logical leap in Volta's presentation. Even if the right combination of metals and moist conductors produces electricity, this does not necessarily imply that torpedoes do not possess a specific electricity of their own. Volta argued by analogy from his pile to the torpedo, maintaining that its internal organs, with their thin disks arranged in columns, even looked like his pile. All the torpedo had to do, he claimed, was to establish electrical contact by squashing its disks together or producing some conductive moisture. As Volta slipped from 'like' to 'is', his torpedoes became electro-motive instruments.

Nevertheless, within a few years, Volta effectively silenced the Galvanic opposition, and the debates about contact and animal electricity were not reopened for another three decades. Benefiting from Napoleon's enthusiastic support, Volta became one of Europe's leading scientific practitioners, showered with academic honours and rewarded with a substantial income. But, like other instruments – the Leyden jar, for instance, or the electrical machine – Volta's pile opened up new fields by being used differently from the way in which it had originally been intended.

In London, the chemist Humphry Davy converted Volta's pile into a powerful analytical instrument for studying the composition of water and other substances. In his public lectures at the recently founded Royal Institution, Davy demonstrated his mastery of this dramatic apparatus. Volta, he declared, had provided 'a key which promises to lay open some of the most mysterious recesses of nature'. Experimenters modified the design of Volta's pile, developing batteries consisting of several troughs, each containing pairs of linked copper and zinc plates immersed in dilute sulphuric acid. Using one of these trough batteries, Davy isolated three elements – sodium, potassium and chlorine – and won the prize that Napoleon had established in Volta's honour. By founding the science of electrochemistry, Davy established himself as the leader of a new type of post-Enlightenment chemistry, a specialised discipline that was conducted by professional experts.[73]

Other experimenters concentrated on the behaviour of what Volta had called the 'endless circulation of the electrical fluid (this *perpetual motion*)' – in other words, electric current. They investigated how different materials affected the current's generation and transmission, and tried to devise mathematical laws describing its operation. During the opening decades of the nineteenth century, researchers all over Europe were investigating the new current electricity, but it was Michael Faraday who was made into the nineteenth-century hero of electrical science.

The son of a blacksmith who became president of the Royal Institution, Faraday came to personify rags-to-riches success stories; in this respect, he is the English Franklin. Fascinated by Davy's public lectures on chlorine, Faraday abandoned his book-binding apprenticeship to become Davy's assistant. Despite his subsequent fame, Faraday was not the first to demonstrate the link between electricity and magnetism: in 1820, it was the Danish lecturer Hans Ørsted who showed his students that an electric current affects the needle of a small compass beneath the wire. Nevertheless, it was Faraday who subsequently developed the laws of electromagnetism, and who invented the electromagnetic motor that now forms an indispensable part of modern life.

In 1931, a best-selling book about Faraday bore an impressive title: *Faraday – the Story of an Errand Boy who Changed the World*. But vital though his achievements were, Faraday relied on the work of his Enlightenment predecessors. He also shared many of their ideals. Like the visionaries of the early Royal Society or the French Revolution, Faraday was concerned to find useful applications for scientific discoveries. A deeply religious man, he believed that the major goal of science was to discover the laws of nature that had been laid down by God. Spurning all honours for himself, Faraday disparaged inventors with mercenary aims, declaring that the pursuit of truth should be a reward in itself.

In retrospect, it seems ironic that such an idealistic scientist should be heralded as the founder of the enormously profitable electromagnetic industry. During a lecture he gave on chlorine, which had recently been isolated, Faraday discussed the objectives of scientific research. Quoting Franklin's famous question, 'What is the good of a newborn baby?', Faraday quipped that for an experimentalist faced with a new discovery, the answer should be: 'Endeavour to make it useful.' According to a wittier (and probably apocryphal) version of this anecdote, the Prime Minister visited Faraday's laboratory to admire his electromagnetic apparatus. 'Of what use is it?', he enquired. 'Why, sir', Faraday supposedly replied, 'there is every probability that you will soon be able to tax it!'.[74]

# 4   Resuscitation

On a cold morning in January 1803, George Forster, a convicted murderer, was hanged at Newgate. After an hour, his body was taken down from the gallows and handed over to a visiting physics professor, who staged an imposing demonstration. Watched by fascinated surgeons, the professor linked the prostrate corpse to a trough battery with metal plates and wires. First he worked on Forster's face. When the current started to flow, the dead jaws quivered and one eye leered open. As the experiment proceeded, Forster's clenched fist rose into the air. Then his legs started to kick violently and his back arched. Some bystanders thought that Forster was being restored to life.

The itinerant experimenter was Giovanni Aldini, Galvani's nephew and one of his most enthusiastic collaborators and propagandists. In the diagrams illustrating Aldini's books, the disembodied hands of invisible experimenters adjust electrical connections to decapitated human corpses; stretched out on slabs, these dissected people resemble his uncle's frogs. When depicted with clinical detachment, Aldini's experiments seem gruesome. But other experimenters had carried out similar trials, and fifteen years later, a Glaswegian professor used a massive battery with 270 pairs of plates on another dead criminal.

Repulsive though his activities may now appear, Aldini was operating at a time when electrical therapy was still in vogue. Enabling the paralysed to stand up and walk did not, perhaps, seem so very different from resuscitating the dead – or those in a state of suspended animation, as Aldini preferred to put it. He was a keen supporter of the Royal Humane Society, whose members prided themselves on their successful resuscitation procedures. By the time that Aldini visited Newgate, administering a shock from an electrical machine was regularly recommended for restoring drowned people to life. Aldini's major innovation was to suggest using a more readily portable pile to treat suffocation when standard methods of revival had failed.

Aldini's recommendation of a pile might seem an act of treachery towards his uncle. But using Volta's pile certainly did not imply adopting his theories or even his name: Aldini always called his apparatus a Galvanic pile, and he advertised his medical therapies as the application of Galvanism. His objectives also reflected Galvani's. Aldini was interested not in investigating a pile's physical and chemical effects, but in exploring the relationships between life, death and electricity. Many men working in the life sciences shared these concerns, and the nature of life became a contentious topic.

During the 'vitalism' controversy of the early nineteenth century, people argued about where to draw the line between living and non-living matter. Extreme materialists denied that there was any fundamental difference between them. On the opposing side, religious believers were horrified by this reductionist approach, insisting that life is a spiritual quality bestowed by God. In 1814, John Abernethy – President of London's Royal College of Surgeons – proposed a compromise. Perhaps, he suggested, life is some sort of 'superadded' element, some 'subtile, mobile, invisible substance'. Could it perhaps, he suggested, be a superfine fluid 'analogous to electricity'? Abernethy's former pupil William Lawrence publicly ridiculed this argument that electricity could stand in for the soul, maintaining that living organisms exhibit special properties because of the way in which the matter composing them is organised. The nature of life, he urged, cannot be discovered by feeble analogies with electricity, but must be sought in matter itself.[75]

These acrimonious debates, widely discussed in journals like the *Edinburgh Review*, were taking place in the years shortly before Mary Shelley wrote *Frankenstein*, which was first published in 1818. From the text of *Frankenstein*, it is clear that she was an avid and knowledgeable reader who was up-to-date with contemporary scientific affairs. Moreover, she was married to Percy Shelley, who only a few years earlier had been enthusing to his friends at Oxford about 'a new engine', envisioning the extra-ordinary effects that might be achieved by a giant galvanic battery 'of colossal magnitude, a well-arranged system of hundreds of metallic plates'.[76]

Mary Shelley had been only five years old when Aldini used Volta's pile to perform his resuscitation experiments, yet her account of Frankenstein's success at Ingolstadt recalls the earlier scene in London:

> *It was on a dreary night of November, that I beheld the accomplishment of my toils. With an anxiety that almost amounted to agony, I collected the instruments of life around me, that I might infuse a spark of being into the lifeless thing that lay at my feet ... by the glimmer of the half-extinguished light, I saw the dull yellow eye of the creature open; it*

*breathed hard, and a convulsive motion agitated its limbs.*[77]

In his preface to *Frankenstein*, Percy Shelley reported that its central theme, bringing a creature to life electrically, 'has been supposed, by Dr Darwin, and some of the physiological writers of Germany, as not of impossible occurrence'.[78] This Dr Erasmus Darwin (whose young grandson Charles later became world famous for his theory of natural selection) was one of the Enlightenment's most eminent physicians, renowned for his own book about evolution, and a friend of Mary Shelley's father. Among his many writings were accounts of medical and biological experiments, which she later explicitly mentioned as a source for *Frankenstein*.

Darwin's most famous book was a long poem called *The Botanic Garden*, whose extensive footnotes contained long yet clear expositions of contemporary scientific ideas and achievements – including electricity. To exonerate himself from accusations of writing frivolous verse, Darwin coined what became a literary slogan by promising 'to inlist Imagination under the banner of Science'.[79] Mary Shelley seems to have taken these words to heart, placing the culmination of Enlightenment electricity at the heart of a book that became one of the most popular works of literature ever written. Enlightenment dreams of technological progress may have been fulfilled, yet *Frankenstein*'s appeal endures because, like Franklin and his fellow optimists, we remain fascinated by the mystery of electricity.

# Further Reading

Cohen, I. Bernard, *Benjamin Franklin's Science*, Cambridge MA: Harvard University Press, 1990. Detailed, enthusiastic account focusing on Franklin's electrical research.

Fortune, Brandon Brame and Deborah J. Warner, *Franklin and his Friends: Portraying the Man of Science in Eighteenth-Century America*, Philadelphia: University of Pennsylvania Press, 1999. Fascinating and lavishly illustrated discussion of scientific portraiture.

Franklin, Benjamin, *The Autobiography and other Writings*, Harmondsworth: Penguin, 1986. Best-selling, endearing and revealing personal narrative.

Hankins, Thomas L., *Science and the Enlightenment*, Cambridge: Cambridge University Press, 1985. Lucid introduction designed for undergraduates.

Heilbron, John L., *Electricity in the Seventeenth and Eighteenth centuries*, Berkeley: University of California Press, 1979. The classic academic study.

Home, Roderick W., 'Introduction', in *Æpinus's Essay on the Theory of Electricity and Magnetism*, eds Roderick Home and P.J. Connor, Princeton NJ: Princeton University Press, 1979, pp. 1–224. A substantial and thorough review of eighteenth-century electricity and magnetism.

Outram, Dorinda, *The Enlightenment*, Cambridge: Cambridge University Press, 1995. The best short modern introduction.

Pera, Marcello, *The Ambiguous Frog: The Galvani–Volta Controversy on Animal Electricity*, Princeton NJ:

Princeton University Press, 1992. Delightfully written, and includes much background material.

Porter, Roy, *Enlightenment: Britain and the Creation of the Modern World*, London: Allen Lane, The Penguin Press, 2000. Wide-sweeping and highly enjoyable survey concentrating on the English Enlightenment.

Rowbottom, Margaret and Susskind, Charles, *Electricity and Medicine: History of their Interaction*, San Francisco: San Francisco Press Inc., 1984. Old-fashioned but informative.

# Notes

1. Hogg, T.J., *The Life of Percy Bysshe Shelley*, London: Edward Moxon, 2 vols, 1858, vol. 1, pp. 62, 70–1.
2. Quotations from Pera, M., *The Ambiguous Frog: the Galvani–Volta Controversy on Animal Electricity*, Princeton: Princeton University Press, 1992, pp. 3–6; and Martin, B., *The Young Gentlemen's and Ladies Philosophy*, London, 2 vols, 1759–63, vol. 1, p. 319.
3. Wesley, J., *Journal*, ed. N. Curnock, London: Charles Kelly, 8 vols, 1909–16, vol. 4, pp. 53–4 (17 February 1753).
4. Fortune, B. and D. Warner, *Franklin and his Friends: Portraying the Man of Science in Eighteenth-Century America*, Philadelphia: University of Pennsylvania Press, 1999, pp. 74–7, 120–1; verse from 'The Rising Glory of America' (1771), by Philip Frenau and Hugh Henry Brackenridge, quoted p. 128.
5. Erasmus Darwin (in 1787), quoted in King-Hele, D., *Erasmus Darwin: A Life of Unequalled Achievement*, London: DLM, 1999, p. 221.
6. Martin, *Young Gentlemen's and Ladies Philosophy*, vol. 1, p. 301.
7. Porter, R., *Enlightenment: Britain and the Creation of the Modern World*, London: Allen Lane, The Penguin Press, 2000, pp. 59–60.
8. In the *Encyclopédie*, quoted in Sutton, G., 'Electric Medicine and Mesmerism', *Isis* 72 (1981), pp. 375–92 (p. 375).
9. Porter, *Enlightenment*, p. 3.

10. Voltaire, F., *Letters on the English,* trans. L. Tancock, London: Penguin, 1980, p. 69.
11. Priestley, J., *The History and Present State of Electricity with Original Experiments*, London, 1767, p. 547.
12. Priestley, J., *Experiments and Observations on Different Kinds of Air*, London: J. Johnson, 1774–7, vol. 1, p. xiv.
13. Priestley, *History and Present State of Electricity*, pp. 548–9.
14. Quoted in Shapin, S., *The Scientific Revolution*, Chicago and London: Chicago University Press, 1996, p. 99.
15. Letter of 1745–6 (possibly by Gowin Knight) to Benjamin Wilson, British Library MSS ADD 30094, f. 16.
16. Rackstrow, B., *Miscellaneous Observations, Together with a Collection of Experiments on Electricity*, London, 1748, p. 50.
17. Letter from John Smeaton to Benjamin Wilson, 24 July 1746, British Library MSS ADD 30094, f. 22.
18. Heilbron, J., *Electricity in the 17th and 18th Centuries*, Berkeley, Los Angeles and London: University of California Press, 1979, p. 313.
19. *Philosophical Transactions* 44, 1746, pp. 211–12 (Johann Winkler).
20. Sigaud de la Fond, J., *Précis Historique et Expérimental des Phénomènes Électriques*, Paris, 1781, pp. 283–92, quotations on pp. 285, 287 (my translation).
21. Cohen, I.B., *Benjamin Franklin's Experiments: a New Edition of Franklin's Experiments and Observations on Electricity*, Cambridge MA: Harvard University Press, 1941, p. 194 (1748 letter to Peter Collinson).
22. 'Animal Magnetism . . . or Count Fig in a Trance', published by W. Price, 2 July 1789.
23. *Philosophical Transactions* 44, 1747, pp. 704–49.
24. Johnson, S., *Rambler* 199, 11 February 1752.
25. Cohen, *Franklin's Experiments*, p. 63 (letter to Collinson).

26. Franklin, B., *The Autobiography and other Writings*, Harmondsworth: Penguin, 1986, pp. 194–5.

27. Cohen, *Franklin's Experiments*, p. 8 (letter of 1783).

28. Franklin, *Autobiography*, p. 106.

29. Peter Collinson, quoted in Cohen, I.B., *Benjamin Franklin's Science*, Cambridge MA: Harvard University Press, 1990, p. 62.

30. Franklin, *Autobiography*, p. 172.

31. Broadsheet, Newport, 1752, reproduced in Fortune and Warner, *Franklin and his Friends*, p. 76.

32. Cohen, *Franklin's Science*, p. 68.

33. Originally in Latin (inscription for a bust of Franklin), by the French minister Anne-Marie-Robert Turgot.

34. Letter to Cadwallader Colden, 12 April 1753, quoted in Cohen, *Franklin's Science*, p. 141.

35. Mitchell, T., 'The Politics of Experiment in the Eighteenth Century: The Pursuit of Audience and the Manipulation of Consensus in the Debate over Lightning Rods', *Eighteenth-Century Studies* 31, 1998, pp. 307–31 (p. 313).

36. Brydone, P., *A Tour through Sicily and Malta*, London, 1776, vol. 1, p. 241.

37. Cohen, *Franklin's Experiments*, p. 62 (1747 letter to Peter Collinson).

38. Mitchell, 'Politics of Experiment,' p. 324.

39. Quoted in Heilbron, *Electricity*, p. 382.

40. Quoted in Cohen, *Franklin's Experiments*, p. 138.

41. Rackstrow, *Miscellaneous Observations*, p. 25.

42. Wesley, J., *The Desideratum: or electricity made plain and useful*, Bristol, 1771, pp. 42–72.

43. Quoted without source on p. 75 of Schaffer, S., 'Self Evidence', in *Questions of Evidence: Proof, Practice, and Persuasion across the Disciplines*, eds. J. Chandler, A. Davidson and H. Harootnian, Chicago: University of Chicago Press, 1994, pp. 56–91.

44. *Philosophical Transactions* 49, 1756, pp. 558–63 (Cheney Hart).

45. Quoted on p. 380 of Sutton, G., 'Electric Medicine and Mesmerism', *Isis* 72, 1981, pp. 375–92.

46. *Philosophical Transactions* 51:1, 1759, p. 358.

47. *Philosophical Transactions* 25, 1706–7, p. 2374.

48. Desaguliers, J.T., *A Course of Experimental Philosophy*, London, 1734, p. vi.

49. Euler, L., *Letters to a German Princess, on Different Subjects in Physics and Philosophy*, London, 1795, vol. 2, pp. 141–2.

50. Hans Sloane, quoted in Heilbron, *Electricity*, p. 252.

51. *Philosophical Transactions* 38, 1734, p. 264.

52. Quoted in Cohen, *Franklin's Experiments*, p. 64.

53. John Cook, quoted on p. 137 of Cantor, G., 'The Theological Significance of Ethers', 1981, in G. Cantor and M. Hodge, *Conceptions of Ether: Studies in the History of Ether Theories, 1740–1900*, Cambridge: Cambridge University Press, 1981, pp. 135–56.

54. Newton, I., *Opticks*, New York: Dover, 1952, p. 352 (4th edition, Query 21).

55. Cantor, 'Theological Significance of Ethers', p. 147.

56. Richard Barton, quoted in Cantor, 'Theological Significance of Ethers', p. 149.

57. Priestley, *History and Present State of Electricity*, p. 732.

58. British Library MSS ADD 30094, f. 91 (undated).

59. Adams, G., *Lectures on Natural and Experimental Philosophy, Considered in it's Present State of Improvement*, London: R. Hindmarsh, 4 vols, 1799, vol. 4, p. 304.

60. *Encyclopædia Britannica*, 1801, vol. 1, p. iv, p. 547 (John Robinson on dynamics).

61. Priestley, *Experiments and Observations on Air*, vol. 2, p. 29.

62. Jungnickel, D. and McCormach, R., *Cavendish: The*

*Experimental Life*, Cranbury NJ: Bucknell University Press, 1999, quotation on p. 307 (Marsillio Landriani).

63. Jungnickel and McCormach, *Cavendish*, p. 131 (John Walker).

64. *Philosophical Transactions* 63, 1773, pp. 476–7.

65. *Philosophical Transactions* 64, 1774, p. 473.

66. Anon, *The Sceptic*, Retford, 1800, pp. 1, 6.

67. Giambattista Beccaria, quoted in Pera, *Ambiguous Frog*, p. 55.

68. Marc'Antonio Caldani, quoted in Pera, *Ambiguous Frog*, p. 54.

69. Galvani, 1782, quoted in Pera, *Ambiguous Frog*, p. 64.

70. Volta, quoted in Pera, *Ambiguous Frog*, p. 175.

71. Volta, A., 'On the Electricity excited by the mere Contact of conducting Substances of different Kinds', *Philosophical Magazine* 7, 1800, pp. 289–337 (p. 305).

72. *Philosophical Magazine* 7, 1800, pp. 289–311.

73. Golinski, J., *Science as Public Culture: Chemistry and Enlightenment in Britain, 1760–1820*, Cambridge: Cambridge University Press, 1992, pp. 188–235 (p. 206).

74. Quotations from pp. 178–9 of Cohen, I.B., 'Faraday and Franklin's "Newborn Baby"', *Proceedings of the American Philosophical Society* 131, 1987, pp. 177–82.

75. Quotations from pp. xviii–xx of Marilyn Butler's introduction to Shelley, M., *Frankenstein*, Oxford and New York: Oxford University Press, 1993.

76. Hogg, *Life of Shelley*, vol. 1, p. 62.

77. Shelley, *Frankenstein*, pp. 38–9.

78. Ibid., p. 3.

79. Darwin, E., *The Botanic Garden*, London: J. Johnson, 1795, p. iii.